Advance Praise for
The Party's Over

Richard Heinberg has distilled complex facts, histories, and events into
_dable overview of the energy systems that keep today's mass
_y running. The result is jarring. *The Party's Over* is the book
we need to reorient ourselves for a realistic future.

– Chellis Glendinning, Ph.D., author of *Chiva: A Village
Takes on the Global Heroin Trade* and *Off the Map:
An Expedition Deep into Empire and the Global Economy*

_v generations hence, our descendants will look back on the
_al world of today with a combination of awe, wonderment, and
_or. Their past is our future — a transitional era of dwindling
_y supplies, resource wars, and industrial collapse. If societies a
_ury from now have managed to learn how to live peacefully,
_destly, and sustainably, it may be at least partly because the
advice in this timely book was heeded.

— Thom Hartmann, author of
The Last Hours of Ancient Sunlight and *Unequal Protection:
The rise of corporate dominance and theft of human rights*

_ichard Heinberg makes shockingly clear in this extraordinarily
_-researched and -written book, our way of life will soon change
_matically, as oil production and reserves both begin to decline.
_o makes clear that our actions now will strongly affect what is
_ft of the world when this shift away from oil takes place.
_ before we can act we must understand, and before we can
_derstand we must be informed. In this compelling book,
_ard Heinberg gives us the tools — the information and
understanding — to act. *The Party's Over* is a
wise and important work.

— Derrick Jensen, author of *A Language Older than Words*
and *The Culture of Make Believe*

The Party's Over begins with a commanding review of world history, where past and current developments including war, empire, and population growth are interpreted as functions of cheap or increasingly scarce and expensive energy. The discussion of substitutes for fast-depleting fossil fuels, and the formidable impediments to making the transition that would allow industrial civilization to continue, are important to every investor and citizen.

— Virginia Deane Abernethy, Ph.D., author of *Population Politics*

Richard Heinberg's *The Party's Over* is outstanding. I hope that the US President and Congress read this book. The world and the US populations are projected to double in 50 and 70 years, respectively, and global oil supplies are projected to be mostly depleted in 50 years! I agree with Heinberg that society is headed for serious trouble in the near future.

— David Pimentel, Professor, Department of Entomology, Systematics and Ecology, Cornell University

Mariners often say that nine tenths of navigation is knowing where you are: Richard Heinberg's *The Party's Over* is the seminal book that locates us most accurately on the dangerous map of industrial life. Heinberg helps lay and expert reader alike to understand oil peak and its staggering ramifications for what many of us consider 'normal life'. The Party's Over provides a solid grounding for grasping both the unfortunate extent of our dependence on the twin hydrocarbons oil and natural gas (and other forms of "big" energy), as well as the enormity of the task of transitioning towards a 'post carbon' world.

— Julian Darley, author of
High Noon for Natural Gas: The New Energy Crisis and coauthor of
Relocalize Now! Getting Ready for Climate Change and the End of Cheap Oil, and founder and director of Post Carbon Institute

Richard Heinberg is absolutely brilliant and more in touch with big-picture issues and small-picture nuances than any writer I know. When Heinberg writes, I listen.

— Michael C. Ruppert, author of *Crossing the Rubicon: The Decline of the American Empire at the End of the Age of Oil,* and publisher of *From the Wilderness*

By the same author:

The Oil Depletion Protocol, A Plan to Avert Oil Wars, Terrorism and Economic Collapse
Powerdown, Options and Actions for a Post-Carbon World

THE PARTY'S OVER

OIL, WAR AND THE FATE OF INDUSTRIAL SOCIETIES

RICHARD HEINBERG

CLAIRVIEW

Clairview Books
Hillside House, The Square
Forest Row, East Sussex
RH18 5ES

www.clairviewbooks.com

Published by Clairview 2003
Reprinted 2004
Second edition 2005
Reprinted 2007

First published in Canada by New Society Publishers, Gabriola Island, 2003

A catalogue record for this book is available from the British Library

ISBN 978 1 905570 00 3

Cover by Andrew Morgan Design
Typeset in Canada
Printed and bound by Cromwell Press Limited, Trowbridge, Wiltshire

Contents

Acknowledgments

I am deeply indebted to three geologists who read parts of the manuscript and offered invaluable corrections, additions, and advice: C. J. Campbell, who read the entire manuscript and offered editorial as well as technical suggestions; Jean Laherrère, who read Chapter 3 and offered detailed criticisms and suggestions; and Walter Youngquist, who read an early version of Chapter 3 and supplied helpful resource materials.

I am also indebted to Ron Swenson, an expert on renewable energies, who offered immensely valuable insights and suggestions for Chapter 4.

For the past several years my students at New College of California have heard me develop the ideas for this book in lectures; during the same period my colleagues on staff and faculty have engaged me in frequent conversations about issues related to the book. I would like to thank all of these wonderful people, both for their comments and for their patience with me as I doggedly pursued this topic.

Readers of my monthly *MuseLetter* read early drafts of chapters and also offered helpful suggestions. I thank them for their loyalty and interest.

Thanks to Chris Plant and the rest of the staff at New Society Publishers for taking on a controversial topic.

I must mention my debt of gratitude to Jay Hanson, whose research and documentation of this topic on his web site <www.dieoff.org> provided much of the original inspiration for this book.

For their kind permission to quote from their work, my appreciation goes out to Michael C. Lynch, Bjørn Lomborg, C. J. Campbell, and Richard C. Duncan.

Finally, I would like to thank my wife, Janet Barocco, for her constant support and encouragement.

Introduction

The skylines lit up at dead of night, the air-conditioning systems cooling empty hotels in the desert, and artificial light in the middle of the day all have something both demented and admirable about them: the mindless luxury of a rich civilization, and yet of a civilization perhaps as scared to see the lights go out as was the hunter in his primitive night.

— Jean Baudrillard (1989)

It is evident that the fortunes of the world's human population, for better or for worse, are inextricably interrelated with the use that is made of energy resources.

— M. King Hubbert (1969)

There is no substitute for energy. The whole edifice of modern society is built upon it .... It is not "just another commodity" but the precondition of all commodities, a basic factor equal with air, water, and earth.

— E. F. Schumacher (1973)

The world is changing before our eyes — dramatically, inevitably, and irreversibly. The change we are seeing is affecting more people, and more profoundly, than any that human beings have ever witnessed. I am not referring to a war or terrorist incident, a stock market crash, or global warming, but to a more fundamental reality that is driving terrorism, war, economic swings, climate change, and more: the discovery and exhaustion of fossil energy resources.

The core message of this book is that industrial civilization is based on the consumption of energy resources that are inherently limited in quantity, and that are about to become scarce. When they do, competition for what remains will trigger dramatic economic and geopolitical events; in the end, it may be impossible for even a single nation to sustain industrialism as we have known it during the twentieth century.

1

What comes after industrialism? It could be a world of lower consumption, lower population, and reduced stress on ecosystems. But the process of getting there from here will not be easy, even if the world's leaders adopt intelligent and cooperative strategies — which they have so far shown little willingness to do. Nevertheless, the end of industrial civilization need not be the end of the world.

This is a message with such vast implications — and one that so contradicts the reassurances we receive daily from politicians and other cultural authorities — that it appears, on first hearing, to be absurd. However, in the chapters that follow I hope to show

- the complete and utter **dependency** of modern industrial societies on fossil fuel energy resources as well as the inability of alternatives to fully substitute for the concentrated, convenient energy source that fossil fuels provide;

- the **vulnerability** of industrial societies to economic and political disruption as a result of even minor reductions in energy resource availability;

- the **inevitability** of fossil fuel depletion;

- the **immediacy** of a peak in fossil fuel production, meaning that soon less will be available with each passing year regardless of how many wild lands are explored or how many wells are drilled;

- the **role of oil** in US foreign policy, terrorism and war, and the geo-politics of the 21st century;

- and hence the necessity of our responding to the coming oil production peak cooperatively, with compassion and intelligence, in a way that minimizes human suffering over the short term and, over the long term, enabling future generations to develop sustainable, materially modest societies that affirm the highest and best qualities of human nature.

I came to the subject of energy resources out of a passion for ecology and a decades-long effort to understand what makes human cultures change — an attempt, that is, to answer the question, *What causes one group of people to live in air-conditioned skyscrapers and shop at supermarkets, while another genetically similar group lives in bark huts and gathers wild foods?*

This is a complex problem. There is no single explanation for the process of cultural change; reasons vary considerably from situation to situation. However, as many students of the subject eventually conclude, there is one element in the process that is surprisingly consistent — and that is the role of energy.

Life itself requires energy. Food is stored energy. Ecosystems organize themselves to use energy as efficiently as possible. And human societies expand or contract, invent new technologies or remain static, in response to available energy supplies. Pay attention to energy, and you can go a long way toward understanding both ecological systems and human social systems, including many of the complexities of economic and political history.

Once I realized this, I began to focus my attention on our society's current energy situation. Clearly, over the past century or so we have created a way of life based on mining and consuming fossil energy resources in vast and increasing quantities. Our food and transportation systems have become utterly dependent on growing supplies of oil, natural gas, and coal. Control of those supplies can therefore determine the economic health and even the survival of nations. Then I tried to find answers to the following questions: *How much petroleum is left? How much coal, natural gas, and uranium? Will we ever run out? When? What will happen when we do? How can we best prepare? Will renewable substitutes — such as wind and solar power — enable industrialism to continue in a recognizable form indefinitely?*

Important questions, these. But a quick initial survey of available answers proved to be confusing and frustrating. There are at least four sets of voices spouting mutually contradictory opinions:

- The loudest and most confident voice belongs to conventional free-market economists, who view energy as merely one priced commodity among many. Like other commodities, energy resources are subject to market forces: temporary shortages serve to raise prices, which in turn stimulates more production or the discovery of substitutes. Thus the more energy we use, the more we'll have! Economics Nobel laureate Robert Solow has gone so far as to say that, ultimately, " ... the world can, in effect, get along without natural resources."[1] Economists like him have a happy, cornucopian view of our energy future. If an energy crisis appears, it will be a temporary one caused by "market imperfections" resulting from government regulation. Solutions will come from the market's natural response to price signals if those signals are not obscured by price caps and other forms of regulatory interference.

- A more strident voice issues from environmental activists, who are worried about the buildup of greenhouse gases in the atmosphere and about various forms of hydrocarbon-based pollution in air, water, and soil. For the most part, ecologists and eco-activists are relatively unconcerned with high energy prices and petroleum resource depletion — which, they assume, will occur too late to prevent serious environmental damage from global warming. Their message: Conserve and switch to renewables for the sake of the environment and our children's and grandchildren's welfare.

- A third and even more sobering collective voice belongs to an informal group of retired and independent petroleum geologists. This is a voice that is so attenuated in the public debate about energy that I was completely unaware of its existence until I began systematically to research the issues. The petroleum geologists have nothing but contempt for economists who, by reducing all resources to dollar prices, effectively obscure real and important physical distinctions. According to the petroleum geologists, this is arrant and dangerous nonsense. Petroleum will run out. Moreover, it will do so much sooner than the economists assume — and substitutes will not be easy to find. The environmentalists, who for the most part accept economists' estimates of petroleum reserves, are, according to the geologists, both right and wrong: we should indeed be switching to renewable alternatives, but because the renewables cannot fully replicate the energy characteristics of fossil fuels and because decades will be required for their full development, a Golden Age of plentiful energy from renewable sources is simply not in the cards. Society must engage in a crash program of truly radical conservation if we are to avoid economic and humanitarian catastrophe as industrialism comes to its inevitable end.

- Finally, there is the voice that really matters: that of politicians, who actually set energy policy. Most politicians tend to believe the economists because the latter's cornucopian message is the most agreeable one — after all, no politician wants to be the bearer of the awful news that our energy-guzzling way of life is waning. However, unlike economists, politicians cannot simply explain immediate or projected energy constraints away as a temporary inconvenience. They have to deal with constituents — voters — who want good news and quick solutions. When office holders are forced to acknowledge the reality of an impending energy crisis, they naturally tend to propose solutions appropriate to their constituency and their political philosophy, and they predictably tend to blame on their political opponents whatever symptoms of the crisis cannot be ignored. Those on the political Left usually favor price caps on energy and subsidies to low-income rate payers; they blame price-gouging corporations for blackouts and high prices. Those on the political Right favor "free-market" solutions (which often entail subsidies to oil companies and privately owned utilities) and say that shortages are due to environmental regulations that prevent companies from further exploration and drilling.

Personally, I have long supported the program of developing renewable energy alternatives that eco-activists advocate. I still believe in that program, now more than ever. However, after studying the data and interviewing experts, I have concluded that, of the four groups described above, the retired and independent petroleum geologists are probably giving us the most useful factual information. Theirs is a long-range view based on physical reality. But their voice is the hardest to hear because, while they have undeniable expertise, there are no powerful institutions helping them spread their message. In this book, the reader will find the geologists' voices prominently represented.

As should be obvious from the title of this book, I am choosing to emphasize the bad news that we are approaching the first stages of an energy crisis that will not easily be solved and that will have a profound and permanent impact on our way of life. There is also good news to be conveyed: it is possible that, in the post-petroleum world, humankind will discover a way of living that is more psychologically fulfilling as well as more ecologically sustainable than the one we have known during the industrial age. However, unless we are willing to hear and accept the bad news first, the good news may never materialize.

Many books published during the past few decades have pleaded with us to reduce our non-renewable energy usage for a variety of reasons — to lessen the greenhouse effect and environmental pollution, to halt the destruction of local communities and cultures, or to preserve human health and sanity. Though I agree with those prescriptions, this is not another such book. Until now, humankind has at least theoretically had a choice regarding the use of fossil fuels — whether to use constantly more and suffer the long-term consequences or to conserve and thus forgo immediate profits and industrial growth. The message here is that *we are about to enter a new era in which, each year, less net energy will be available to humankind, regardless of our efforts or choices.* The only significant choice we will have will be how to adjust to this new regime. That choice — not whether, but how to reduce energy usage and make a transition to renewable alternatives — will have profound ethical and political implications. But we will not be in a position to navigate wisely through these rapids of cultural change if we are still living with the mistaken belief that we are somehow entitled to endless energy and that, if there is suddenly less to go around, it must be because "they" (the Arabs, the Venezuelans, the Canadians, the environmentalists, the oil companies, the politicians, take your pick) are keeping it from us.

Industrial societies have been flourishing for roughly 150 years now, using fossil energy resources to build far-flung trade empires, to fuel the invention of spectacular new technologies, and to fund a way of life that is opulent and fast-paced. It is as if part of the human race has been given a sudden windfall of wealth and decided to spend that wealth by throwing an extravagant party. The party has not been without its discontents or costs. From time to time, a lone voice issuing from here or there has called for the party to quiet down or cease altogether. The partiers have paid no attention. But soon the party itself will be a fading memory — not because anyone decided to heed the voice of moderation, but because the wine and food are gone and the harsh light of morning has come.

Here is a brief tour of the book's contents:

Chapter 1 is a general discussion of energy in nature and human societies. In it we see just how central a role energy has played in the past and why it will shape the fates of nations in the decades ahead. This chapter is a brief guided trip through the fields of ecology, cultural anthropology, and history, with energy as our tour guide.

Chapter 2 traces the history of the industrial era — the historic interval of cheap energy — from the Europeans' first use of coal in the 12th century to the

20th-century miracles of petroleum and electricity with their cascading streams of inventions and conveniences.

Chapter 3 is in many respects the informational core of the book. In it we will learn to assess oil resources and review estimates of current reserves and extraction rates. Many readers may find the information in this chapter unfamiliar and disturbing since it conflicts with what we frequently hear from economists and politicians. Among other things, we will explore the question, *Why do the petroleum-reserve estimates of independent geologists diverge so far from those of governmental agencies like the US Geological Survey?*

Chapter 4 explores the available alternatives to oil: from coal and natural gas to solar power, wind, and hydrogen, including cold fusion and "fringe" free-energy devices.

Chapter 5 discusses the meaning and the implications of the approaching peak in fossil-fuel production. We will explore the connections between petroleum dependence, world food systems, and the global economy. We will also examine the global strategic competition for dwindling petroleum resources and attempt to predict the flashpoints for possible resource wars.

Finally, Chapter 6 addresses the vital question: *What can we do?* — individually, as communities, as a nation, and globally. In this chapter we will explore solutions, from the simple practical steps any of us can take to policy recommendations for world leaders. As we will see, humankind now must decide whether to respond to resource shortages with bitter competition or with a spirit of cooperation. We will face this decision at all levels of society — from the family and neighborhood to the global arena of nations and cultures.

Energy, Nature and Society

The life contest is primarily a competition for available energy.

— Ludwig Boltzman (1886)

Other factors remaining constant, culture evolves as the amount of energy harnessed per capita per year is increased, or as the efficiency of the instrumental means of putting the energy to work is increased. We may now sketch the history of cultural development from this standpoint.

— Leslie White (1949)

[T]he ability to control energy, whether it be making wood fires or building power plants, is a prerequisite for civilization.

— Isaac Asimov (1991)

We live in a universe pulsing with energy; however, only a limited amount of that energy is available for our use. We humans have recently discovered a temporary energy subsidy in the forms of coal, oil, and natural gas, and that momentary energy bonanza has fueled the creation of modern industrial societies. We tend to take that subsidy for granted, but can no longer afford to do so. Emerging circumstances will require us to think much more clearly, critically, and contextually about energy than we have ever done before.

In this chapter we will first review some basic facts about energy and the ways in which nature and human societies function in relation to it. We will follow this discussion of principles with an exploration of the history of the United States' rise to global power, showing the central role of energy resources in that process.

The first section below includes information that may already be familiar to many readers from high-school or college courses in physics, chemistry, and biology. I begin with this material because it is absolutely essential to the understanding of all that follows throughout the book. Have patience. We will soon arrive in new (and disturbing) intellectual territory.

Energy and Earth: The Rules of the Game

Few understand exactly what energy is. And yet we know that it exists; indeed, without it, *nothing* would exist.

We commonly use the word *energy* in at least two ways. A literary or music critic might say that a particular poem or performance has energy, meaning that it has a dynamic quality. Similarly, we might remark that a puppy or a toddler has a lot of energy. In those cases we would be using the term intuitively, impressionistically, even mystically — though not incorrectly. Physicists and engineers use the word to more practical effect. They have found ways to measure energy quite precisely in terms of ergs, watts, calories, and joules. Still, physicists have no more insight into energy's ultimate essence than do poets or philosophers. They therefore define energy not in terms of what it is, but by what it does: as "the ability to do work" or "the capacity to move or change matter." It is this quantifiable meaning of the term *energy* that concerns us in this book. Though we are considering something inherently elusive (we cannot, after all, hold a jar of pure energy in our hands or describe its shape or color), energy is nevertheless a demonstrable reality. Without energy, nothing happens.

In the 19ᵗʰ century, physicists formulated two fundamental laws of energy that appear to be true for all times and places. These are commonly known as the First and Second Laws of Thermodynamics. The first, known as the Conservation Law, states that energy cannot be created or destroyed, only transformed. However, energy is never actually "transformed" in the sense that its fundamental nature is changed. It is more accurate to think of energy as a singular reality that manifests itself in various forms — nuclear, mechanical, chemical, thermal, electromagnetic, and gravitational — which can be converted from one to another.

The Second Law of Thermodynamics states that whenever energy is converted from one form to another, at least some of it is dissipated, typically as heat. Though that dissipated energy still exists, it is now diffuse and scattered, and thus less available. If we could gather it up and re-concentrate it, it could still work for us; but the act of re-concentrating it would itself require more energy. Thus, in effect, available energy is always being lost. The Second Law is known as the law of entropy — a term coined by the German physicist Rudolf Clausius in 1868 as a measure of the amount of energy no longer practically capable of conversion into work. The Second Law tells us that the entropy within an isolated system inevitably increases over time. Since it takes work to create and maintain order within a system, the entropy law tells us that, in the battle between order and chaos, it is chaos that ultimately will win.

It is easy to think of examples of entropy. Anyone who makes the effort to keep a house clean or who tries keeping an old car repaired and on the road knows about entropy. It takes work — thus energy — to keep chaos at bay. However, it is also easy to think of examples in which order seems naturally to increase. Living things are incredibly complex, and they manage not only to maintain themselves but to produce offspring as well; technological gadgets (such as computers) are always becoming more sophisticated and capable; and human societies seem to become larger, more complex, and more powerful over time. These phenomena all appear to violate the law of entropy. The key to seeing why they actually don't lies in the study of systems.

The Second Law states that it is the entropy in an *isolated system* that will always increase. An isolated system is one that exchanges no energy or matter with its environment. The only truly isolated system that we know of is the universe. But there are two other possible types of energy systems: *closed systems* (they exchange energy with their environment, but not matter) and *open systems* (they exchange both energy and matter with their environment). The Earth is, for the most part, a closed system: it receives energy from the Sun and re-radiates much of that energy back out into space; however, aside from the absorption of an occasional asteroid or comet fragment, the Earth exchanges comparatively little matter with its cosmic environment. Living organisms, on the other hand, are examples of open systems: they constantly receive both energy and matter from their environment, and also give off both energy and matter.

It is because living things are open systems, with energy and matter continually flowing through them, that they can afford to create and sustain order. Take away their sources of usable energy or matter, and they soon die and begin to disintegrate. This is also true of human societies and technologies: they are open systems that depend upon the flow of energy and matter to create temporary islands of order. Take away a society's energy sources, and "progress"

— advances in technology and the growth of complex institutions — quickly ceases. Living systems can increase their level of order and complexity by increasing their energy flow-through; but by doing so, they also inevitably increase the entropy within the larger system of which they are a part.

Matter is capable of storing energy through its chemical order and complexity. This stored energy can be released through chemical processes, such as combustion or, in the case of living things, digestion. Materials that store energy are called *fuels*.

The law of entropy holds true for matter as well as for energy. When energy is dissipated, the result is called *heat death*. When matter is eroded or degraded, the result is called *matter chaos*. In both cases, the result is a randomization that makes both matter and energy less available and useful.

In past decades, a simplistic understanding of entropy led many scientists to conclude that order is an anomaly in the universe — a belief that made it difficult to explain how biological evolution has proceeded from the simple to the complex, from bacteria to baleen whales. In recent years, more sophisticated understandings have developed, centered mostly around chaos theory and Ilya Prigogine's theory of dissipative structures. Now it is known that, even within apparently chaotic systems, deeper forms of order may lurk. However, none of these advances in the understanding of living systems and the nature of entropy circumvents the First or Second Laws of Thermodynamics. Order always has an energy cost.

Because the Earth is a closed system, its matter is subject to entropy and is thus continually being degraded. Even though the planet constantly receives energy from its environment, and even though the ecosystems within it recycle materials as efficiently as they can, useful concentrations of matter (such as metal ores) are always being dispersed and made unusable.

On Earth, nearly all the energy available to fuel life comes from the Sun. There are a very few exceptions; for example, oceanographers have discovered organisms living deep in ocean trenches, thriving on heat emanating from the Earth's core. But when we consider the energy flows that support the biosphere as a whole, sources originating within the planet itself are trivial.

The Sun continually gives off an almost unimaginable amount of energy — the equivalent of roughly 100 billion hydrogen bombs going off each second —radiating it in all directions into space. The Earth, 93 million miles away, is a comparatively tiny target for that energy, receiving only an infinitesimal fraction of what our local star radiates. Still, in terms that concern us, that's plenty:

our planet is constantly bathed in 1,372 watts of sunlight energy per square meter. The total influx of solar energy to the Earth is more than 10,000 times the total amount of energy humankind presently derives from fossil fuels, hydro power, and nuclear power combined. The relative vastness of this solar-energy influx as compared with society's energy needs might suggest that humans will never face a true energy shortage. But only some of this solar energy is actually available for our use: much is re-radiated into space (30 percent is immediately reflected from clouds and ice), and nearly all of the rest is already doing important work, such as driving the weather by heating the atmosphere and oceans and fueling life throughout the biosphere.

Some organisms — green plants, including algae and phytoplankton — are able to take in energy directly from sunlight. Biologists call these organisms *producers*, or *autotrophs* ("self-feeders"), because they make their own food from inorganic compounds in their environments.[1] Producers trap solar energy through photosynthesis, a process in which chlorophyll molecules convert sunlight into chemical energy. Most of us tend to assume that green plants are mostly made up of materials from the soil drawn up through the plants' roots. This is only partly true: plants do require minerals from the soil, but most of their mass is actually derived from air, water and sunlight, via photosynthesis. Hundreds of chemical changes are involved in this process, the results of which can be summarized as follows:

$$\text{carbon dioxide} + \text{water} + \text{solar energy} \longrightarrow \text{glucose} + \text{oxygen}$$
$$6CO_2 + 6H_2O + \text{solar energy} \longrightarrow C_6H_{12}O_6 + 6O_2$$

Glucose — a sugar, or carbohydrate — serves as food for plants and can be converted into materials from which the plants build their tissues. Plants absorb only about half of the solar energy that falls on them; of that, they are able to convert only about one to five percent into chemical energy. Still, even at this low level of efficiency, photosynthetic organisms each year capture a little more than twice the total amount of energy used annually by human beings. (However, within the US, the total amount of energy captured in photosynthesis amounts to only about half of the energy used by humans.)

All nonproducing organisms are classifiable as *consumers*, or *heterotrophs* ("other-feeders"). By digesting glucose and other complex organic compounds that were produced through photosynthesis, consumers absorb the energy previously locked into chemical order by green plants. In the process, they produce waste — less-ordered material — which they excrete into the environment. In effect, consumers feed on order and excrete chaos in order to survive. All animals are consumers.

There are several categories of consumers: *herbivores*, which eat plants; *carnivores*, which eat other consumers (primary carnivores eat herbivores, secondary carnivores eat other carnivores, and tertiary carnivores eat carnivores that eat carnivores); *scavengers*, which eat dead organisms that were killed by other organisms or died naturally; *detritovores*, which eat cast-off fragments and wastes of living organisms; and *decomposers*, consisting mostly of certain kinds of bacteria and fungi, which complete the final breakdown and recycling of the remains and wastes of all organisms. Human beings — like foxes, bears, rats, pigs, and cockroaches — are *omnivores*, eating both plants and animals.[2]

Both producers and consumers use the chemical energy stored in glucose and other organic compounds to fuel their life processes. In most cells, this is accomplished through aerobic respiration, a process with a net chemical change opposite that of photosynthesis:

glucose + oxygen $\longrightarrow$ carbon dioxide + water + energy
$$C_6H_{12}O_6 + 6O_2 \longrightarrow 6CO_2 + 6H_2O + \text{energy}$$

Some decomposers get energy through anaerobic respiration, or fermentation. Instead of carbon dioxide and water, the end products are compounds such as methane gas (a simple hydrocarbon) and ethyl alcohol. Normally, in the decay of organic materials, a chemical process based on aerobic respiration occurs, with carbon-based organic material combining with oxygen to yield carbon dioxide and water. However, if there is no additional oxygen available because of an anaerobic environment — such as exists if organic matter is buried under sediment or stagnant water — then anaerobic decomposers go to work. Plant and animal remains are transformed into hydrocarbons as oxygen atoms are removed from the carbohydrate organic matter. This is the chemical basis for the formation of fossil fuels. It is now believed that most oil comes from a few brief epochs of extreme global warming over quite short spans of geological time. The process began long ago and today yields fuels — chemically stored sunlight — that are energy-dense and highly usable.

Energy in Ecosystems: Eating and Being Eaten

Just as individual organisms use energy, so do complex systems made up of thousands or millions of organisms. The understanding of how they do so has been one of the central projects of the science of ecology.

The term *ecology* was coined in 1869 by German biologist Ernst Haeckel from the Greek roots *oikos* ("house" or "dwelling") and *logos* ("word" or "study of"). However, the discipline of ecology — which is the study of how organisms interact with one another and their surroundings — did not really flourish until the beginning of the 20ᵗʰ century.

At first, ecologists studied food chains — big fish eating little fish. Quickly, however, they realized that since big fish die and are subsequently eaten by scavengers and microbes that are then eaten by still other organisms, it is more appropriate to speak of food *cycles* or *webs*. Further analysis yielded the insight that all of nature is continually engaged in the cycling and recycling of matter and energy. There are carbon cycles, nitrogen cycles, phosphorus cycles, sulfur cycles, and water cycles. Of fundamental importance, however, are *energy flows* — which tend to drive matter cycles and which, as we have seen, begin in nearly all cases with sunlight.

Energy is the basic currency of ecosystems, passing from green plants to herbivores to carnivores, with decomposers participating along the way. With each transfer of energy, some is lost to the environment as low-quality heat. Typically, when a caterpillar eats a leaf, when a thrush eats the caterpillar, or when a hawk eats the thrush, only 5 to 20 percent of usable energy is transferred from one level to the next. Thus, if green plants in a given area capture, for example, 10,000 units of solar energy, then roughly 1,000 units will be available to support herbivores, even if they eat all of the plants; only 100 units will be available to support primary carnivores; only 10 to support secondary carnivores; and only one to support tertiary carnivores. The more energy-transfer levels there are in the system, the greater the cumulative energy losses. In every ecosystem, most of the chemically bound energy is contained among the producers, which also account for most of the *biomass*. The herbivores present will account for a much smaller fraction of the biomass, and the carnivores for yet a still smaller fraction. Thus the energy flow in ecosystems is typically represented by a pyramid, with producers on the bottom and tertiary carnivores at the top.

The energy available in an ecosystem is one of the most important factors in determining its *carrying capacity*, that is the maximum population load of any given species that is able to be supported by its environment on an ongoing basis. Energy is not the only factor, however; the operative principle in determining carrying capacity is known as Liebig's Law (after the 19ᵗʰ-century German scientist Justus von Liebig), which states that whatever necessity is least abundant, relative to per-capita requirements, sets the environment's limit for the population of any given species. For a plant, the limiting factor may be heat, sunlight, water, nitrogen, or phosphorus. Sometimes too much of a limiting factor restricts the carrying capacity, as when plants are killed by too much water or too much soil acidity. The limiting factor for any population may change over time. For herbivores and carnivores, the most common limiting factor is food-energy. This is why ecologists pay so much attention to food webs: when we understand the energy flows within an ecosystem, the dynamics of the system as a whole become clear.

These days the term *ecology* is often understood to be used merely in a scientific critique of human society's negative impact on nature. There are two reasons for this. The first is that early ecologists soon realized that, since humans are organisms, ecology should include the study of the relationship between people and the rest of the biosphere. The second is that, as early ecologists cataloged and monitored various natural systems, they found that it was becoming increasingly difficult to study such systems in an undisturbed state; everywhere, nature was being impacted by the human presence.

This impact itself became a focus of investigation, and soon ecologists realized that disturbed and undisturbed systems differ in clear ways. Ecosystems that have not been disturbed significantly for long periods of time (whether by humans or by natural disasters) tend to reach a state of dynamic equilibrium which ecologists call a *climax phase*, meaning that organisms have adapted themselves to one another in such a way as to maintain relatively constant population levels, to avoid direct competition, to keep energy flow-through to a minimum, and to recycle available energy and nutrients as completely as possible. They have formed, to use an anthropomorphic term, a *community*.

Biological communities are kept in equilibrium through balancing *feedback loops*. A useful technological example of a balancing feedback loop is a thermostat: if a room gets too cold, the thermostat triggers the furnace to turn on; when the room achieves the set temperature, the thermostat turns the furnace off. The temperature of the room varies, but only narrowly. Similarly, feedback loops in ecosystems — such as predator-prey relationships — tend to keep varying population levels within narrow ranges. If the vole population increases, fox and hawk populations will soon expand to take advantage of this food-energy surplus. The increase in the hawk and fox populations will then reduce the vole population, whose diminution will eventually lead to a reduction in the numbers of hawks and foxes as well.

The more mature the ecosystem, the more thoroughly the organisms in it use the available energy. Waste from one organism becomes food for another. Moreover, in order not to expend energy unnecessarily, organisms will tend to avoid direct competition through any of several strategies: by dividing the habitat into niches, by specializing (for example, if two species depend upon the same food source, they may evolve to feed at different times of day), or by periodic migration. Territorial animals avoid wasting energy in fights by learning to predict one another's behavior from signals like posture, vocalizations, and scent marks.[3] As a result, climax ecosystems give the appearance of cooperation and harmony among member species. The degree of mutual interdependence achieved can be astounding, with differing species relying on one another for food, shelter, transportation, warnings of danger, cleaning, or protection from predators. As biologist Lewis Thomas once put it, "The urge to form partnerships, to link up in collaborative arrangements, is perhaps the oldest, strongest, and most fundamental force in Nature. There are no solitary, free-living creatures, every form of life is dependent on other forms."[4]

In climax ecosystems, population levels are kept relatively in check not only through predators culling prey species, but also through species acting on their own to limit their numbers via internal feedback mechanisms. These internal mechanisms are seen in elephants, for example, which regulate their population densities through delays in the onset of maturity as well as among smaller animals such as mice, where females typically ovulate more slowly or cease ovulation altogether if populations become too dense. In many bird species, much of the adult population simply does not breed when there is no food-energy available to support population growth.

All of this contrasts with ecosystems that have recently been seriously disturbed, or whose balances have been upset by the arrival of a new species.

Fires, floods, and earthquakes are high-energy events that can overwhelm the energy balances of climax ecosystems. Disturbed ecosystems are characterized by disequilibrium and change. First, *pioneer species* appear — and proliferate wildly. They then give way to various secondary species. The environment passes through a series of phases, known collectively as *ecological succession*, until it arrives again at a climax phase. During these successive phases, earlier organisms transform the environment so that conditions are favorable for organisms that appear later. For example, after a forest fire, tough, annual, weedy, ground-cover plants spring up first. During the second or third season, perennial shrubs begin to dominate; a few years later, young trees will have grown tall enough to shade out the shrubs. In some cases, this first generation of trees may eventually be replaced by other tree species that grow taller. It may take many decades or even centuries for the land to again become a climax forest ecosystem. If we accept the view that the Earth can itself be treated as a living being, as has been proposed by biologists James Lovelock and Lynn Margulis[5], then it might be appropriate to think of succession as the Earth's method of healing its wounded surface.

In other instances, balances in ecosystems can be upset as a result of the appearance of *exotic species*. These days, the arrival of most exotic species is due to the actions of humans importing plants and animals for food, decoration, or as pets. But sometimes new arrivals appear on a freak wind current or a piece of flotsam. Most newcomers, having evolved in other environments, are unfit for life in their new surroundings and quickly perish; but occasionally, an exotic species finds itself in an environment with plenty of available food and with no predators to limit its numbers. In such instances, the species becomes an *invader* or *colonizer* and can compete directly with indigenous species. Most Americans are familiar with Scotch broom, starlings, and kudzu vine — all of which are successful, persistent, and profuse colonizers.

Many colonizing species are parasites or disease-causing organisms: bacteria, protozoa, or viruses. When such organisms initially invade a host species, they are often especially virulent because the host has not yet developed the proper antibodies to ward off infection. But the death of the host is no more in the interest of the microbe than it is in the interest of the host itself since the former is dependent on the latter for food and habitat. Thus, over time, disease organisms and their hosts typically co-evolve, so that diseases which initially were fatal eventually become relatively innocuous childhood diseases like measles, mumps, or chickenpox.

Not all feedback loops create balance, however; in *reinforcing feedback* loops, change in one direction causes more change in the same direction. A technological example would be a microphone held too close to the speaker of the amplifier to which it is attached. The microphone picks up sound coming from the speaker, then feeds it back to the amplifier, which amplifies the sound and sends it back through the speaker, and so on. The result is a loud, unpleasant squeal.

Colonizing species sometimes create reinforcing feedback loops within natural systems. While population levels among species in climax ecosystems are relatively balanced and stable, populations in disturbed or colonized ecosystems go through dramatic swings. When there is lots of food-energy available to the colonizing species, its population *blooms*. Suppose the organism in question is the rabbit, and the environment is Australia — a place previously devoid of rabbits, where there is plenty of food and no natural predator capable of restraining rabbit population growth. Each rabbit adds (on average) ten new baby rabbits to the population. This means that if we began with ten rabbits, we will soon have 110. Each of these adds ten more, and before we know it, we have 1,210 rabbits. More rabbits cause more babies, which cause more rabbits, which cause more babies.

Obviously, this cannot go on forever. The food supply for the rabbits is ultimately limited, and eventually there will be more rabbits than there is food to support them. Over the long term, a balance will be struck between rabbits and food. However, that balance may take a while to be achieved. The momentum of population increase may lead the rabbits to *overshoot* their carrying capacity. The likelihood of overshoot is increased by the fact that the environment's carrying capacity for rabbits is not static. Since the proliferating rabbits may eat available vegetation at a faster rate than it can naturally be regenerated, the rabbits may actually reduce their environment's rabbit-carrying capacity even as their numbers are still increasing. If this occurs, the rabbit population will not simply gradually diminish until balance is achieved; instead, it will rapidly *crash* — that is, the rabbits will *die off.*

At this point, depending on how seriously the rabbits have altered their environment's carrying capacity, they will either adapt or die out altogether. If they have not eaten available food plants to the point that those plants can no longer survive and reproduce, the rabbit population will stabilize at a lower level. For a time, population levels will undergo more seasonal swings of bloom, overshoot, and die-off as food plants recover and are again eaten back. Typically, those swings will slowly diminish as a balance is achieved and as the rabbits become incorporated into the ecosystem. This is, in fact, what has begun to happen in Australia since the introduction of rabbits by Europeans in 1859. However, if the rabbits were ever to eat food plants to the point of total elimination, they would reduce the rabbit-carrying capacity of their environment to zero. At that point, the rabbits would die out altogether.

Since successful invaders change their environments, usually overpopulating their surroundings and overshooting their ecosystem's carrying capacity, colonized ecosystems are typically characterized by reduced diversity and increased energy flow-through. As colonizers proliferate, energy that would ordinarily be intercepted by other organisms and passed on through the food web goes unused. But this is always a temporary state of affairs: living systems don't like to see energy go to waste, and sooner or later some species will evolve or arrive on the scene to use whatever energy is available.

These are the rules of the game with regard to energy and life: energy supplies are always limited; there is no free ride. In the long run, it is in every species' interest to learn to use energy frugally. Competition, though it certainly exists in Nature, is temporary and limited; Nature prefers stable arrangements that entail self-limitation, recycling, and cooperation. Energy subsidies (resulting from the disturbance of existing environments or the colonization of new ones) and the ensuing population blooms provide giddy moments of extravagance for some species, but crashes and die-offs usually follow. Balance eventually returns.

Social Leveraging Strategies: How to Gain an Energy Subsidy

We don't often tend to think about the social sciences (history, economics, and politics) as subcategories of ecology. But since people are organisms, it is apparent that we must first understand the principles of ecology if we are to make sense of events in the human world.

Anthropological data confirm that humans are capable of living in balance and harmony as long-term members of climax ecosystems. For most of our existence as a species, we survived by gathering wild plants and hunting wild animals. We lived within the energy balance of climax ecosystems — altering our environment (as every species does), yet maintaining homeostatic, reciprocally limiting relationships with both our prey and our predators.

However, humans are also capable of acting as colonizers, dominating and disrupting the ecosystems they encounter. And there is evidence that we began to do this many millennia ago, long before Europeans set out deliberately to colonize the rest of the world.

Like all organisms, humans seek to capture solar energy. Humans have certain disadvantages as well as advantages in this regard. Our disadvantages include our lack of thick fur, which would allow us to live in a wide range of climates, and our upright posture, which hampers our ability to outrun bears and lions. Our advantages include our adaptability, our flexible and grasping hands, and our ability to communicate abstract ideas by means of complex vocalizations — that is, by language.

We have made the most of our advantages. By exploiting them in ever more ingenious ways, we have developed five important strategies for gaining energy subsidies and thereby expanding the human carrying capacity of our environments:

- *takeover,*
- *tool use,*
- *specialization,*
- *scope enlargement,* and
- *drawdown.*[6]

While other creatures have adopted some of these strategies to a limited degree, modern industrial humans have become masters of all of them, combining and leveraging their advantages. Through an examination of these strategies we can begin to understand how and why *Homo sapiens* — one species among millions — has come to dominate the planetary biosphere.

Takeover

The first and most basic strategy that we have used to increase the human carrying capacity of our environments is one that William Catton, in his pathbreaking book *Overshoot* (1980), called *takeover*. It consists, in his words,

> ... of diverting some fraction of the earth's life-supporting capacity from supporting other kinds of life to supporting our kind. Our pre-*Sapiens* ancestors, with their simple stone tools and fire, took over for human use organic materials that would otherwise have been consumed by insects, carnivores, or bacteria. From about 10,000 years ago, our earliest horticulturalist ancestors began taking over *land* upon which to grow crops for human consumption. That land would otherwise have supported trees, shrubs, or wild grasses, and all the animals dependent thereon — but fewer humans. As the expanding generations replaced each other, *Homo sapiens* took over more and more of the surface of this planet, essentially at the expense of its other inhabitants.[7]

Takeover is a strategy composed of substrategies. The most basic of these entailed simply moving to new habitats. *Homo sapiens* presumably evolved in Africa; probably because of population pressure (which, in turn, may have been due to natural disasters or climate change), early humans left their African homeland and gradually began to fan out around the globe — first to Asia and Europe, and then to Australia, the Pacific Islands, and the Americas. As humans arrived in new habitats, they inevitably took over food-energy from other organisms, as all successful colonizing species do. They hunted for wild game that might otherwise have been prey for wolves, lions, or bears; and they foraged for roots, berries, seeds, and tubers that were already nourishment to a host of herbivores.

Meanwhile, humans were themselves prey to large carnivores. Hence, humans and the existing members of their newfound ecosystem communities went through a process of mutual adjustment. The archaeological evidence suggests that the adjustment was sometimes a painful one: humans often upset local balances dramatically, appropriating so much of the food supply that they caused or hastened the extinction of many animal species.[8]

Humans facilitated the takeover process by the use of fire — a rapid release of chemically stored energy. This constituted a second substrategy of takeover. In addition to keeping people warm at night, fire also served to increase their food supply. Early humans often carried fire sticks with them, deliberately igniting underbrush both to flush out game and to encourage the growth of edible shoots and grasses. The Native Americans and Aborignals of Australia were still using fire this way when European colonists first arrived. It is interesting to note that at least one nonhuman animal has adopted the same tactic: the black kite of India is known as the "fire hawk" because of its habit of picking up smoldering sticks from fires, dropping them on dry grass, and then waiting to catch small animals that flee.[9]

When humans arrived in Australia roughly 60,000 years ago, their use of fire so disrupted the normal growth cycles of shrubs and trees that large indigenous birds and mammals, including giant kangaroos and flightless ostrich-like birds, were deprived of food. According to recent paleontological research, roughly 85 percent of the Australian animals weighing more than 100 pounds disappeared within a few millennia of the first human appearance on the scene.[10]

The first humans to arrive in the Americas and the Pacific Islands provide similar examples: there, too, animal extinctions closely followed human arrival. In North America, the mammoth, mastodon, native horse, four-pronged antelope, native camel, giant beaver, ground sloth, mountain deer, and giant peccary all succumbed about 12,000 to 10,000 years ago, at a time when humans were migrating rapidly from Asia through present-day Alaska and southward into vast territories opened up by retreating ice sheets. Similarly, the Polynesian peoples extinguished the large, flightless moa bird soon after arriving in New Zealand.

But it is important to note what happened next in many of these places. In ancient Australia, over a period of tens of thousands of years, human beings and their adopted environment achieved a relative balance. The Aboriginals developed myths, rites, and taboos: overhunting was forbidden, and burning was permitted only in certain seasons of the year. Meanwhile, native species adjusted themselves to the presence of humans. All of the surviving species — humans, animals, and plants — co-evolved. By the time European colonizers arrived, once again upsetting the balance, Australia — people and all — had the characteristics of a climax ecosystem. Many native Australian trees and shrubs had so adjusted themselves to the Aboriginals' "fire-farming" practices that they could no longer reproduce properly in the absence of deliberate burning. Moreover, the Aboriginals had learned the necessity of limiting their own population levels through extended lactation, the use of contraceptive herbs, or, if necessary, infanticide.

In North America, native peoples had come to regard as sacred the animals and plants they used as food. According to Luther Standing Bear in his 1928 book *My People the Sioux*, Native Americans recognized a human responsibility to the rest of nature and regarded "the four-leggeds, the wingeds, the star people of the heavens, and all things as relatives."[11] Overhunting or the wanton destruction of ecosystems had come to be viewed by these people as an act with negative moral as well as practical implications.

In addition to the colonization of new territories and the use of fire, humans have pursued takeover through yet another substrategy: the appropriation of ever greater amounts of the total food web to human use, first through horticulture (gardening with a hoe or digging stick), then through agriculture (the planting of field crops, usually entailing the use of plows and draft animals). The deliberate planting and tending of food plants probably began gradually and somewhat inadvertently at a time when humans had already populated many habitable areas of the world as densely as they could. When people live by hunting and gathering, they require large territories; in this case, the human carrying capacity of a typical environment may be considerably less than one person per square mile. Horticulture yielded more food from a given land area, permitting population densities of several individuals per square mile.

Agriculture was yet more productive, permitting even greater population densities, though it also resulted in a reduction in the variety and nutritional quality of the human food supply: paleoanthropologists have found that the skeletons of early agriculturalists are usually smaller and show more evidence of degenerative diseases than those of earlier hunter-gatherers.

Agriculture entailed the deliberate simplification of ecosystems. Humans learned to grow only a few domesticated food crops while discouraging competitors to their food plants (weeds) and killing any organisms that competed with humans for access to those food plants (pests).

The domestication of animals constituted yet another variation on the takeover strategy. Animals could be useful for extracting energy from ecosystems in two ways: first, by concentrating and making available food energy from otherwise inedible fibrous plants; and second, by providing traction to pull plows, carts, and carriages. By helping to intensify agricultural production and assisting in overland transportation, domesticated animals facilitated the conquest of ecosystems and continents.

Though the takeover strategy was applied at first to other species, soon some humans began to use it in relation to other humans. Typically, societies with denser populations and more powerful weapons took over the territories of, or enslaved, groups with less intensive demands on the environment. This last substrategy achieved its apotheosis in the European takeover of most of the rest of the planet throughout the past 500 years.

Tool Use

Over the millennia, we humans facilitated our takeover of new ecosystems and other societies with an expanding kit of tools — from fire-drills, spears, knives, baskets, and pots to plows, carts, sailboats, machine guns, steam shovels, and computers.

This second basic strategy — the design, making, and use of tools — has ancient roots: archaeological evidence suggests that humans have been using tools for at least a hundred thousand years, perhaps much longer. Moreover, tool use is not absent among other animals: captive birds of the corvid family (which includes crows, ravens, and jays) have been reliably observed spontaneously constructing rakes out of available sticks or newspaper strips for pulling grain from outside their cage; placing stones in a drinking dish to raise the water to a drinkable level; or using a plastic cup to fetch and pour water on too-dry food.[12] Thus, the spectacular tools invented and used by modern industrial humans represent the development of a long-existing biological potential.

Nearly all tools assist in the harvesting or leveraging of ever-greater amounts of energy from the environment. The only notable exceptions are tools used purely for entertainment — which are also ancient, dating back at least to the oldest-recovered bone flute, made about 60,000 years ago.

It is often said that humans use tools to adapt and change their environments, and this is certainly true (recall the use of fire to thin out brush and thus clear space for the growth of food-yielding plants). However, it is just as accurate to say that we use tools to adapt *ourselves* to a variety of habitats. For example, we use shoes to adapt our feet to walking on rocky or uneven terrain.

Looked at this way, tools can be considered as functionally equivalent to detachable organs.[13] Another way of saying this is that tools are *prosthetic devices* we add to ourselves to replace or supplement our senses, limbs, or muscles. Usually the term *prosthesis* is used to describe a mechanical replacement for an absent organ or a supplement for a poorly functioning one (examples include artificial limbs, false teeth, iron lungs, and eyeglasses); however, it is possible to broaden the concept to include mechanical enhancements of perfectly healthy organs: wheels enhancing the mobility of legs and feet, bows and arrows effectively extending the reach of arms and hands, and so on. William Catton calls *Homo sapiens* "the prosthetic animal" and notes wryly that "when an airline pilot with thirty-three years of flying experience refers to the familiar act of buckling his cockpit seatbelt as 'strapping a DC-8 to my waist,' it is clear that even a modern jetliner can be seen as an elaborate prosthetic device."[14] Catton also notes that the "evolutionary and ecological significance of such prosthetic devices has been to facilitate the spread of mankind over a more extensive range than we could have occupied with only the equipment of our own bodies."[15]

Because tools are extensions of ourselves, they change us. The human-tool complex is effectively a different organism from a toolless human. We unconsciously tend to adapt ourselves to our tools in a myriad of ways — witness how industrial societies have adapted themselves to the automobile. Tool use also alters the mentality of entire societies. For example, the use of the technology of money tends to move whole cultures in the direction of an increased emphasis on calculation and quantification, powerfully intensifying any existing utilitarian attitudes toward natural resources and other humans by facilitating the accumulation of wealth. Similarly, as Marshall McLuhan and others have documented, the technology of writing reduces people's reliance upon memory while intensifying their use of abstract reasoning.[16] More recently, computers have sped up our lives while seeding our language with new metaphors: we now "process" experiences the way our computers process information; we get together with friends to "download" gossip; we complain that talkative individuals take up too much "bandwidth"; we go on vacations so that we can have "down time." Gone are the days of barnyard metaphors (chickens coming home to roost, foxes guarding the henhouse, grown children leaving the nest). As metaphors based on experiences of the natural world disappear from language and are replaced by mechanical or electronic referents, human consciousness may be subtly disengaging itself from its biological roots.

One way to better understand the evolution of technology through the millennia is to examine the relationship between tools and energy. All tools require energy for their use or manufacture — but that energy may come from human muscle power or some source external to the human body, such as animal muscle, wood fire, coal fire, or hydro-generated electricity. Some tools harness externally produced energy, making it available to other tools that then do work for us. Using energy source as a criterion, we can identify four basic categories of tools. These categories also correspond very roughly to four major watersheds in social evolution:

A. *Tools that require only human energy for their manufacture and use.* Examples include stone spearheads and arrowheads, grinding tools, baskets, and animal-skin clothing. These sorts of tools are found in all hunter-gatherer societies.

B. *Tools that require an external power source for their manufacture, but human power for their use.* Examples: all basic metal tools, such as knives, metal armor, and coins. These tools were the basis of the early agricultural civilizations centered in Mesopotamia, China, Egypt, and Rome.

C. *Tools that require only human energy for their manufacture, but harness an external energy source.* Examples: the wooden plow drawn by draft animals, the sailboat, the firedrill, the windmill, the water mill. The firedrill was used by hunter-gatherers, and the wooden plow and sailboat were developed in early agricultural societies; the windmill and water mill appeared at later stages of social evolution.

D. *Tools that require an external energy source for their manufacture and also harness or use an external energy source.* Examples: the steel plow, the gun, the steam engine, the internal combustion engine, the jet engine, the nuclear reactor, the hydroelectric turbine, the photovoltaic panel, the wind turbine, and all electrical devices. These tools and tool systems are the foundation of modern industrial societies —in fact, they define them.

This scheme of classification emphasizes the cumulative nature of technological and social development. Some Class A tools still persist in horticultural, agricultural, and even industrial societies (flint blades, for example, are, because of their extreme sharpness, today often used by brain and eye surgeons for the most delicate operations), but Class D tools by and large did not exist in hunter-gatherer societies. However, the categories do overlap somewhat, and there are exceptions and anomalies: hunter-gatherers used fire to make some tools (for example, by cooking glues), thus turning them into Class C tools; the use of the metal plow (Class D) predated industrialism by three millennia; and a simple steam engine (Class D) was invented by the ancient Greeks, though they did not put it to practical use. Still, even if we allow for these inconsistencies, the scheme shows a clear trend: over time, tools and the societies that use them have increasingly captured energy from sources external to the human body and used that captured energy to fashion even more sophisticated energy-capturing and energy-reliant tools and tool systems.

Specialization

This third strategy is closely related to the second. Since a human-tool complex is effectively a different organism from a toolless human, humans using different tool complexes can become, in effect, different species from one another. As a society becomes composed of people working in different occupations, using different sets of tools, it becomes more complex; it develops its own technological-economic "ecosystem" that exists within, yet apart from, the larger biotic ecosystem.

We noted earlier that humans first applied the takeover strategy to other species and then to other humans; something similar happened with the tool-using strategy. At first, humans made tools out of stones and sticks, but eventually their increasingly utilitarian frame of mind led them to begin treating other human beings as tools. This scheme at first took the form of slavery. Some humans could capture the energy of others who had been seized in war, putting them to work at tasks too dangerous, dreary, or physically taxing for any free person to undertake voluntarily — tasks such as mining metal ores from beneath the Earth's surface. Those ores were, in turn, the raw materials from which were fashioned the chains and weapons that kept the slaves themselves in bondage. Eventually, metals also came to be used as money, a tool that would become the basis for a more subtle form of energy capture: wage labor. Through the payment of money, humans could be persuaded to give their energies to tasks organized by — and primarily benefiting — others. Some humans would become members of a permanent soldier class, which, through its conquests, could capture human slave-energy; others would become part of a peasant class, capturing solar energy through the growing of plants and animals for food for others. Compared to the raw energy of fire, human energy is of extremely high quality because it is intelligently directed. Only with the computer revolution of the late 20ᵗʰ century could inventors envision automatons capable of capturing and using energy in comparably sophisticated ways.

Just as the use of tools has affected our collective psychology, so has special-ization. With a lifelong division of labor, many members of society became cut off from basic subsistence activities and processes; rather than enjoying a direct relationship with the natural world, they became, for their material existence, dependent upon the society's economic distribution system. This subtly fos-tered attitudes of conformity and subordination while undermining feelings of personal confidence and competence.

Scope Enlargement

To understand the nature of this fourth strategy for enlarging the human car-rying capacity of environments, we must return to Liebig's Law, which states that for any given organism the carrying capacity of a region is limited by whatever indispensable substance or circumstance is in shortest supply.

Tools provided ways of getting around many limiting factors. For example, clothing permitted humans to live in climates that were otherwise too cold, whereas irrigation enabled humans to produce an abundance of food in regions that would otherwise have supported far fewer inhabitants. However, some limiting factors could be mitigated simply by transporting resources from one region to another. This sharing of resources among geographically circum-scribed regions typically took the form of trade.

If one region had plenty of minerals but poor soil and another had good soil but no minerals, trade allowed both regions to prosper so that the total pop-ulation of the two regions working together could far exceed what would be possible if they remained in isolation. William Catton calls this strategy *scope enlargement* and argues that

> a good many of the events of human history can be seen as efforts
> to implement [this principle] .... Progress in transport technology,
> together with advancements in the organization of commerce, often
> achieved only after conquest or political consolidation, have had the
> effect of enlarging the world's human carrying capacity by enabling

more and more local populations (or their lifestyles) to be limited
not by local scarcity, but by abundance at a distance.[17]

Local or regional catastrophes — famines, earthquakes, floods, droughts,
plagues, etc. — have always been part of the human experience. With scope
enlargement, their effects can be somewhat offset, as when aid is trucked or
flown into a region experiencing famine. However, local populations then tend
to become increasingly dependent on the system of trade and transport that
connects them. If that system were itself ever to be threatened, many or all of
the regions it encompasses would suddenly be put at risk.

In the past few decades, the strategy of scope enlargement has reached its
logical culmination in a world system of trade and transport known as *globalization*.
We who today live in industrialized countries are the ultimate heirs of the
millennia-long process of scope enlargement. We have become globalized humans,
daily eating foods grown hundreds or thousands of miles away, filling our cars
with gasoline that may have originated in oil wells on the other side of the planet.

Drawdown

The fifth and final strategy that humans have used to increase their environment's
carrying capacity is to find and draw down nature's stocks of nonrenewable
energy resources: coal, oil, natural gas, and uranium. This strategy can only be
pursued once societies are near the point of being able to invent, and produce
in quantity, sophisticated Class D tools.

Drawdown dramatically improved the rates of return from the previous four
strategies. It permitted

- the intensification of agriculture, with chemical fertilizers, pesticides,
  and herbicides increasing yields per acre, and with acreages devoted to
  the growing of food for humans increasing as a result of draft animals
  being replaced by tractors;
- the invention and utilization of a vast array of new tools that use energy
  more intensively;
- the development of more social roles and occupations based on specialized
  tool usage; and
- the rapid acceleration of transportation and trade.

Drawdown has been by far the most successful of the five strategies at increasing
the human carrying capacity of the planet, and the degree of that success can be
gauged in a single statistic, namely that of the world population growth since

the beginning of the industrial revolution. The human population did not reach one billion until about 1820; in the less than two centuries since then, it has increased nearly six-fold. This is a rate of growth unprecedented in human history.

The exploitation of energy-bearing minerals created so much new carrying capacity, and so quickly, that much of that new capacity could be translated into increased wealth and a higher standard of living for a small but significant portion of the world's population. Previously, a parasitic increase of the standard of living for a wealthy few (kings, nobles, and lords) nearly always entailed a lessening of the standard of living of far more numerous serfs and peasants. Now, with power being liberated from fossil fuels, so much energy was available that the standard of living could be improved for large numbers of people, at least to a certain extent. Even though the majority of the world's population shared but little in this bonanza and continued to be exploited for cheap labor via takeover and specialization, virtually everyone shared in the expectation that the benefits of fuel-fed industrialism could eventually be spread to all. This expectation led in turn to a partial relaxation of the class-based social tensions that had plagued complex societies since their beginnings.

Americans, more than the people of any other region, have learned to take high-energy living standards for granted. In order to gain some perspective on this accustomed standard, it might be helpful to perform a little experiment. Try running up three flights of stairs in twenty seconds. If you weigh 150 pounds and the three flights go up forty feet, you will have done 6,000 foot-pounds of work in twenty seconds, or 300 foot-pounds per second. One horsepower equals 550 foot-pounds per second; therefore, you will have just generated a little over half a horsepower. But no one could sustain such a burst of muscle-energy all day long. The average sustained human power output is roughly one-twentieth of a horsepower.

This exercise is useful (even if performed only in imagination) in comparing human power with the power of the machines that maintain our modern way of life. Suppose human beings were powering a generator connected to one 150-watt light bulb. It would take five people's continuous work to keep the light burning. A 100-horsepower automobile cruising down the highway does the work of 2,000 people. If we were to add together the power of all of the fuel-fed machines that we rely on to light and heat our homes, transport us, and otherwise keep us in the style to which we have become accustomed, and then compare that total with the amount of power that can be generated by the human body, we would find that each American has the equivalent of over 150 "energy slaves" working for us 24 hours each day. In energy terms, each middle-class American is living a lifestyle so lavish as to make nearly any sultan or potentate in history swoon with envy.[18]

But if the payoffs of the drawdown strategy are spectacular, so are its dangers and liabilities. The latter can be grouped into three broad categories:

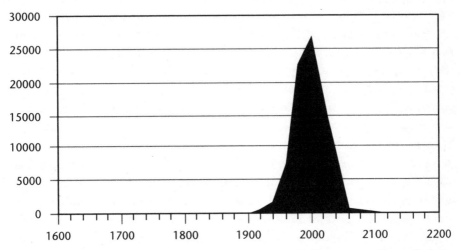

Figure 1. World oil production from 1600 to 2200, history and projection, in millions of barrels per year (Source: C. J. Campbell)

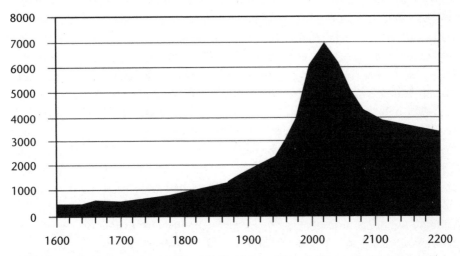

Figure 2. World population from 1600 to 2200, history and projection, assuming impacts from oil depletion, in millions (Source: C. J. Campbell)

environmental degradation, climate change, and increasing human dependency on a "phantom" carrying capacity.

Pollution was the first drawback of fossil fuel use to make itself apparent. Of course, pollution was hardly unknown before fossil fuels — it was apparent in the smoke of wood fires blackening winter skies over medieval cities, the horse manure clogging streets in 19th-century London and New York, and the tailings from mines ruining surrounding land and water throughout most of the civilized world since the dawn of civilization itself. But with the advent of the petrochemical industry, the toxic load on the environment has increased dramatically and quickly. Over the course of a few decades, chemical engineers synthesized tens of thousands of new, complex organic compounds for a wide variety of purposes. Few of these chemicals were safety-tested; of those that were, many turned out to have toxic effects on humans or other organisms. The undesirable consequences of the spread of these chemicals into the environment were sometimes dramatic, with rates of respiratory ailments and cancers soaring, and at other times more subtle, with estrogen-mimicking chemicals disrupting reproductive processes in fish, birds, amphibians, and mammals, including humans.[19]

The second danger of the drawdown method, which has more recently begun to make itself known, is climate change resulting from the global accumulation of greenhouse gases. The world's oil and coal fields represent vast stores of carbon that have been sequestered under the Earth's surface for hundreds of millions of years. With the advent of the industrial revolution, as these stores of carbon began to be mined and burned at an increasing rate, that carbon was released into the atmosphere as carbon dioxide (CO_2). There is strong evidence to suggest that elevated levels of carbon dioxide trap heat in the global atmosphere, creating a greenhouse effect that gradually warms the planet. Climate records derived from Greenland ice cores indicate a very close correlation between atmospheric carbon dioxide concentrations and global temperatures. Around the beginning of the 20th century, both CO_2 concentrations and global temperature began perceptibly to rise. For the previous 10,000 years, the amount of carbon in our atmosphere had remained constant at 280 parts per million. By 1998, that amount had increased to 360 ppm and was projected to increase to 560 ppm by the middle of the current century. Climate scientists have projected a consequent increase in the average global temperature of 3 to 7 degrees Fahrenheit (2 to 5 degrees Celsius).

Thus we have, unintentionally, begun to disturb massive planetary systems that have kept much of the world climate relatively hospitable to civilization for the last 10,000 years. We are heating the deep oceans, which leads to more frequent and intense El Niño weather patterns. The timing of the seasons is noticeably altered and most of Earth's glaciers are retreating at accelerating rates. The potential effects are catastrophic. They include the drowning of coastal cities and whole island nations as a result of rising sea levels and intensified storms; the proliferation of disease-spreading insects into new regions, resulting in cases of malaria perhaps doubling in tropical regions and increasing 100-fold elsewhere; and the loss of forests and wildlife that depend upon a stable climate, leading to vastly increased extinction rates and the collapse of whole ecosystems.[20] The Earth's climate is so finely balanced that global warming could result in a rapid flip in weather regimes. For example, cold, fresh water from the melting of the arctic ice pack could halt the Gulf Stream, plunging Europe and North America into a new Ice Age.

The third danger of the drawdown strategy is one that is discussed less frequently than either pollution or global warming, though its ultimate implications for humankind may be even more dire. This is our increasing dependency on energy resources that are depleting within historically narrow time frames. There are now somewhere between two and five billion humans alive who probably would not exist but for fossil fuels. Thus if the availability of these fuels were to decline significantly without our having found effective replacements to maintain all their life-sustaining benefits, then the global human carrying capacity would plummet — perhaps even below its pre-industrial levels. When the flow of fuels begins to diminish, everyone might actually be worse off than they would have been had those fuels never been discovered because our pre-industrial survival skills will have been lost and there will be an intense competition for food and water among members of the now-unsupportable population (Chapter 5 provides a closer look at the likely consequences of the anticipated petroleum depletion.).

Complexity and Collapse: Societies in Energy Deficit

The five strategies humans have adopted for capturing increasing amounts of energy (takeover, tool use, specialization, scope enlargement, and drawdown) have permitted societies to grow in size, scope, and complexity. However, it is important to note that the ramp of history, rising upward from the simplest Paleolithic hunter-gatherer bands to the heights of globalized industrial civilization, has not been a smooth one. Many civilizations have expanded their scope and complexity dramatically, only to dissolve back into simpler forms of social organization.

Archaeologists have understandably given much attention to the study of collapsed complex societies since the ruins left by the ancient Egyptians, Romans, Mayas, Greeks, Minoans, Mesopotamians, Harappans, and Chacoans provide a wealth of material for investigation. Why would a group of people intelligent enough to have built impressive temples, roads, and cities suddenly lose the ability to maintain them? Why would a society capable of organizing itself into a far-flung empire, with communications networks and distribution systems, suddenly lose its ability to continue? Such questions — as much as the ruins left behind — contribute to a widespread and perennial fascination with lost civilizations.

The literature on the subject is voluminous and includes speculation on the causes of collapse ranging from class conflict to mismanagement. Undoubtedly, the best modern research on this subject was done by archaeologist Joseph Tainter, whose book *The Collapse of Complex Societies* (1988) is now widely recognized as the standard work on the topic. In his book and related essays, Tainter takes an ecological view of society as an energy-processing structure and concludes that complex societies tend to collapse because *their strategies for energy capture are subject to the law of diminishing returns.*

Tainter describes complexity as a problem-solving strategy used by civilizations and empires. "For the past 12,000 years," he writes, these societies "have seemed almost inexorably to grow more complex. For the most part this has been successful: complexity confers advantages, and one of the reasons for our success as a species has been our ability to increase rapidly the complexity of our behavior."[21]

When Tainter uses the term "complexity," he is referring to "such things as the size of a society, the number and distinctiveness of its parts, the variety of specialized roles that it incorporates, the number of distinct social personalities present, and the variety of mechanisms for organizing these into a coherent, functioning whole."[22] Hunter-gatherer societies, for example, may have no more than a few dozen distinct social personalities whereas a modern census recognizes many thousands of occupational roles. More complex societies, Tainter notes,

> are more costly to maintain than simpler ones, requiring greater support levels per capita. As societies increase in complexity, more networks are created among individuals, more hierarchical controls are created to regulate these networks, more information is processed,

there is more centralization of information flow, there is increasing
need to support specialists not directly involved in resource production, and the like. All of this complexity is dependent upon energy
flow at a scale vastly greater than that characterizing small groups of
self-sufficient foragers or agriculturalists. The result is that as a society evolves toward greater complexity, the support costs levied on
each individual will also rise, so that the population as a whole must
allocate increasing portions of its energy budget to maintaining
organizational institutions. This is an immutable fact of societal evolution, and is not mitigated by type of energy source.[23]

Tainter offers the following diagram (Fig.3) as a schematic representation
of the trajectory of a typical complex society. At first, incremental investments
in social complexity, new technologies, and expanding scope yield impressive
returns. Agricultural production increases, and wealth captured from conquest
flows freely as the society's increasingly formidable army invades surrounding
states. But gradually the rates of return tend to diminish, even as requirements
for further investments in institutional support (including investments in legit-
imization and coercion) are still increasing. This eventually makes the strategy
of complexity itself less palatable to the population. According to Tainter,

> a society that has reached this point cannot simply rest on its accom-
> plishments, that is, attempt to maintain its marginal return at the
> status quo, without further deterioration. Complexity is a problem-
> solving strategy. The problems with which the universe can confront
> any society are, for practical purposes, infinite in number and end-
> less in variety. As stresses necessarily arise, new organizational and
> economic solutions must be developed, typically at increasing cost
> and declining marginal return. The marginal return on investment
> in complexity accordingly deteriorates, at first gradually, then with
> accelerating force. At this point, a complex society reaches the phase
> where it becomes increasingly vulnerable to collapse.[24]

From the perspective of the average citizen, the burden of taxes and other
costs is increasing while at the local level there are fewer benefits. The idea of
being independent thus becomes more and more attractive. Collapse, then,
may simply entail the decomposition of society, as individuals or groups decide
to pursue their own immediate needs rather than the long-term goals of the
leadership. In other situations, collapse may entail the takeover of a society that

is stressed because of declining marginal returns by another society that is still enjoying higher rates of return on its investments in strategic leveraging.

Tainter discusses this theory in relation to the well-documented collapse of 17 different civilizations. Regarding the Roman Empire, he writes:

> The establishment of the Roman Empire produced an extraordinary return on investment, as the accumulated surpluses of the Mediterranean and adjacent lands were appropriated by the conquerors. Yet as the booty of new conquests ceased, Rome had to undertake administrative and garrisoning costs that lasted centuries. As the marginal return on investment in empire declined, major stress surges appeared that could scarcely be contained with yearly Imperial budgets. The Roman Empire made itself attractive to barbarian incursions merely by the fact of its existence. Dealing with stress surges required taxation and economic malfeasance so heavy that the productive capacity of the support population deteriorated. Weakening of the support base gave rise to further barbarian successes, so that very high investment in complexity yielded few benefits superior to collapse. In the later Empire the marginal return on investment in complexity was so low that the barbarian kingdoms began to seem preferable.[25]

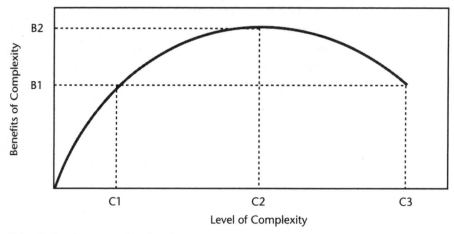

Figure 3: Benefits to a society from investments in complexity over time. Prior to point C1/B1, benefits are abundant; between points B1/C1 and B2/C2, returns on investments in complexity gradually diminish. After a society passes point B2/C2, its returns on investment become negative and it becomes vulnerable to collapse.

(Source: Joseph Tainter, "Complexity, Problem Solving, and Sustainable Societies")

This process of collapse is somewhat analogous to the phenomenon of population overshoot and die-off within a colonized ecosystem; indeed, the population of the city of Rome declined from over a million inhabitants in 100 AD to about 40,000 in 1100 AD.

Tainter's discussion of the Western Chou Empire, the Harappan Civilization, Mesopotamia, the Egyptian Old Kingdom, the Hittite Empire, the Classic Mayan civilization, and others shows a similarly tight fit between theory and historical data.

Western civilization from the Middle Ages to the present illustrates the theory in a somewhat different way. Rather than growing and declining in a simple curve, Western civilization has recovered and undergone at least two even greater growth surges due to its ability to find and exploit new energy subsidies at critical moments. The takeover of the Americas, Africa, India, and the Pacific Islands offered subsidies ranging from slave labor to new sources of metal ores and timber. The expansion of the Euro-American cultural and political influence that these new resources enabled, while impressive, probably could not have been sustained through the 20th century in the face of rising costs (e.g., for the maintenance of colonial administrations) and declining returns, had it not been for the discovery of fossil fuels, the greatest energy subsidy ever known. This discovery, as we have already seen, enabled the transformation of civilization itself into a form never before seen: industrialism.

The returns on early investments in drawdown and industrial production were staggering. Costs were extraordinary as well, but they could easily be borne. As Tainter puts it,

> with subsidies of inexpensive fossil fuels, for a long time many consequences of industrialism effectively did not matter. Industrial societies could afford them. When energy costs are met easily and painlessly, the benefit/cost ratio of social investments can be substantially ignored (as it has been in contemporary industrial agriculture). Fossil fuels made industrialism, and all that flowed from it (such as science, transportation, medicine, employment, consumerism, high-technology war, and contemporary political organization) a system of problem solving that was sustainable for several generations.[26]

This does not mean, however, that industrial civilization is immune to the law of diminishing returns. Tainter cites statistics indicating that already there have been steep reductions in returns on increasing US investments in education, military hardware, information processing, and scientific research. As we will see in more detail in Chapter 3, the drawdown of fossil fuels is itself subject to the law of diminishing returns. Early investments in drilling for oil yielded fabulous returns. But most of the largest and most productive oil fields were discovered within a century of the drilling of the first commercial well: rates of discovery peaked in the 1960s. And so, over time, the amount of energy that must be expended to find and extract each barrel of oil, or to mine each ton of coal, increases.

Tainter ends his book by drawing the following sobering conclusion: "However much we like to think of ourselves as something special in world history, in fact industrial societies are subject to the same principles that caused earlier societies to collapse."[27]

Applied Socio-Ecohistory: Explaining the American Success Story

So far in this chapter we have explored some of the basic energy principles at work in natural systems and human societies. In order to better illustrate these principles (and especially those discussed in the last two sections), let us use what we have learned to address a specific question that could add importantly to our understanding of global energy resource usage over the past two centuries: *Why is the United States of America currently the wealthiest and most powerful nation in the history of the world?*

Often this question is addressed through a discussion of ideas, personalities, and unique historical occurrences. We have all learned the names of early explorers, inventors, and politicians; we have been taught the importance of the American system of government, with its guarantees of freedoms and rights; and we have memorized the dates of important wars and other political events in US history. These are all of course essential to any explanation of US ascendancy. However, let us take an approach that focuses on energy and tackle the following question: To what extent does America owe its prominent position in the world to energy resources and its people's ability to exploit them?

Such a discussion must begin with geology and geography. The North American continent, which Europeans began to explore and claim in the early 16th century, was a place of extraordinary biotic and mineral abundance. Early Spanish conquistadors found vast forests, animals for food and fur, fertile farmland, fresh water, iron, copper, silver, and gold — all in far greater quantities than existed in Europe. Eventually, the colonists' descendants also found an abundance of coal and petroleum. These energy resources proved to be especially valuable because they enabled the more intensive extraction and use of all other resources.

When Europeans first arrived in the New World, there were already other humans present. Why hadn't Native Americans taken more advantage of all these resources? Why was it not they who became world conquerors, sailing to Europe to claim it as a possession of the Iroquois, the Seminole, or the Lakota?

As Jared Diamond explains in his Pulitzer Prize-winning book *Guns, Germs, and Steel: The Fates of Human Societies*, Eurasia had been blessed with indigenous domesticable cereal grains and traction animals nonexistent in the Americas.[28] These permitted — perhaps even encouraged — the development of large-scale agriculture and stratified societies. The Europeans thus had a head start in applying the leveraging strategies discussed above. Their successes in expanding the carrying capacity of their environment meant that Europe, by the 16th century, was comparatively crowded and resource-depleted. Europeans were therefore highly motivated to expand their application of the takeover and scope-enlargement strategies by conquering and exploiting new lands. Most Europeans who came to America were not so much searching for freedom as escaping population pressure and resource depletion.

Still, things might have turned out differently: in the early 15th century, squadrons of large Chinese junks made several amazing voyages that carried them as far as Hormuz; had these expeditions continued, the Chinese might have become the first to circumnavigate Africa and sail the Atlantic and the Pacific. However, political troubles back home in China called a halt to the entire project; thus newly claimed territories in America acquired names like New Spain and New England, rather than New Beijing or New Canton.

As it turned out, the Europeans who arrived in North America regarded the land as essentially empty and saw the native peoples — who were making far fewer demands on resources than the Europeans themselves were accustomed to making — as unproductive savages. Europeans at first sought to enslave the natives, thus taking over the human muscle-energy of the continent in addition to its other resources. But many of the natives — millions, in fact; in some regions over 90 percent of the population — quickly succumbed to colonists' diseases, such as smallpox, measles, and influenza. These diseases were caused by microorganisms that had become integrated into the internal bodily ecosystems of Europeans through centuries of contact with domesticated animals; for the natives of the Americas, however, these microorganisms were exotic invasive species whose impact was utterly devastating.[29] In any case, the natives made poor slaves because most were accustomed to living in a more easy-going and egalitarian — namely less specialized and complex — social environment than were the Europeans, and often preferred death to lifelong servitude.

Nevertheless, it was clear that great wealth could be extracted from the continent if only there were sufficient energy available to farm the land and mine the ores. Quickly, Europeans seized upon the strategy of importing Africans as slaves. With the latter's intelligently directed muscle-power as motive force, the machinery of extraction went to work and produced great fortunes for thousands of colonists and their families — those, that is, who could afford to buy into this wealth-producing system. Because the Africans were typically kidnapped from kingdoms — complex societies — and then ripped from their cultural matrix (not only by transplanting them geographically but by preventing them from speaking their own languages and engaging in their own customs), they were somewhat more easily enslaved than were most Native Americans.

This discussion of "where" and "who" helps account for America's meteoric rise from colonial backwater to global superpower in a mere two centuries, but it is still not sufficient. We must also take into account the "when" of the US appearance on the world scene. Europeans had in fact arrived in North America several centuries before Columbus: the Norse and possibly the Irish made the voyage repeatedly between approximately 1000 and 1350 AD. However, all that ultimately resulted was the leaving behind of a few enigmatic stone inscriptions for future historians to puzzle over. As every musician knows, timing is of the essence. Jared Diamond notes that the

> second Eurasian attempt to colonize the Americas [in the 15th century] succeeded because it involved a source, target, latitude, and time that allowed Europe's potential advantages to be exerted. Spain, unlike Norway, was rich and populous enough to support exploration and subsidize colonies. Spanish landfalls in the Americas were at subtropical latitudes highly suitable for food production, based at first mostly on Native American crops but also on Eurasian domestic animals, especially cattle and horses. Spain's transatlantic colonial enterprise began in 1492, at the end of a century of rapid

development of European oceangoing [Class C] ship technology, which by then incorporated advances in navigation, sails, and ship design developed by Old World societies (Islam, India, China, and Indonesia) in the Indian Ocean.[30]

Resources are of little benefit without the ability to exploit them. Imagine having several barrels of gasoline but no car or other motorized equipment with which to put that gasoline to use. This was essentially the situation not only of the Native Americans, but also, at first, of the invading Europeans with regard to America's energy minerals. Though the continent was rich in coal and petroleum, few people, if any, yet realized that fact.

However, the Europeans had spent many centuries making prior investments in tool making, and so the breakthrough to the production of Class D tools was for them merely the next step in a long evolution of strategic leveraging. As we have already noted, the entire process of industrialization was based on using fossil fuels (initially coal, later petroleum) to mechanize production and transport. Soon after the Industrial Revolution began in England, it became clear that North America in fact had a much greater natural abundance of energy minerals than did Europe. If the US had remained a colony, its energy resources would likely have been siphoned off to promote the production of still more wealth in the Old World. However, the American Revolutionary War had dissolved the former Crown Corporations of Virginia, Delaware, Massachusetts, etc., so that the people of the new nation of the United States of America were free to shape their own economic destiny by exploiting the continent's resources for their own benefit. Thus within a few decades the situation changed from being one in which Europe was taking resources from North America to one in which North America was taking industrial technology from Europe and putting it to more effective use due to its richer resource base. The US did not start the Industrial Revolution, but was poised to capitalize on it.

The history of the 19th century in America is a tale of snowballing invention, exploration, and extraction, each feeding the others. Political events were largely shaped by resource disputes. For example, the realization (by the industrial northern states) that America's future wealth lay far more in the extraction and use of concentrated fuels than in the continued reliance (by the agrarian southern states) on kidnapped African muscle-power may have played a role in the freeing of the slaves.

Overall, the US made the most of its energy-resource advantage. At first, wood fueled the mills and factories of the Northeast; soon it also fueled the railroads that brought raw materials to the factories and manufactured goods to the frontier. In the latter decades of the 19th century, coal took the place of dwindling wood supplies; and then in the 20th, oil — flowing initially from Pennsylvania and Ohio, then from southern California, then Texas and Oklahoma, and finally the Gulf of Mexico and Alaska — in turn fueled the automobile industry, modern agriculture, and the modern chemical industry. While European nations had to colonize far-off places like Indonesia in order to fill their increasing appetite for energy resources, the US could extract all it needed from within its borders. Its energy-resource base was so great that, until 1943, it remained a net petroleum exporter.

In the 20th century, while the old colonial powers (such as England, Spain, and Portugal) were reaping diminishing returns from their investments in conquest and while other aspiring colonial powers (Germany, Japan, and Italy) were thwarted in gaining access to energy resources in other lands, the US found itself in the rare and enviable position of having both abundant indigenous resources and the expertise, technology, and freedom to exploit them for its own benefit. It invested the wealth from these resources both in further technological development and in the production of by far the most powerful and sophisticated weapons systems the world has ever seen. Thus by the end of the Second World War the US was, from both an economic and a military point of view, the most powerful nation in the history of the world.

This is not to say that the promise of political and religious freedom had played no role in drawing millions of skilled and highly motivated immigrants from Europe — though many were simply driven out by overcrowding at home. Nor can one deny the role of extraordinary personalities: inventors, politicians, military leaders, and explorers whose names and accomplishments fill history books. However, it is also indisputable that without its wealth of minerals and energy resources, the US could never have achieved its current position of global dominance.

But American resources, however vast, were nevertheless limited. Throughout the 20th century, geologists combed the North American continent for oil, coal, and natural gas reserves. The US quickly became the most explored region of the planet. Americans were encouraged through advertising to buy private automobiles in order to take advantage of these energy resources, and they did so at a rate unparalleled in the industrialized world. By mid-century, however, older oil wells were running dry and newer wells were proving to be less productive. The rate of discovery of new petroleum resources in the continental US peaked in the 1930s; the rate of extraction of those resources peaked in 1970. But the energy-based "American Way of Life" had to be maintained in order to avoid political and economic disaster; therefore, further energy resources had to come from elsewhere.

Understandably, industrial and political leaders adopted a time-tested strategy — scope enlargement, or trade and transport — in order to make up the difference. The US began to buy oil at first, and soon natural gas, from other nations. Its balance of trade — historically positive — soon became overwhelmingly negative. Formerly the world's foremost lender and investor, the US soon became the world's foremost debtor nation. Meanwhile it continued to develop its already awesome military capability with which to enforce its priorities on the rest of the world, more blatantly so following the demise of its only competitor for global hegemony: the Soviet Union, itself geologically blessed with energy resources but handicapped by early barriers in exploiting those resources and by an economic-social system that discouraged individual initiative.

Soon after US petroleum production had peaked, official policy began emphasizing "free trade" as a global panacea for unemployment, underdevelopment, despotism, and virtually every other economic or political ill. Through its manipulation of the rules of global trade, the US sought to maintain and increase its access to natural resources worldwide. Those rules — written primarily by US-based corporations and encoded in policies of the International Monetary Fund (IMF), the World Bank, and the World Trade Organization (WTO) as well as in treaties like the North American Free Trade Agreement (NAFTA) — essentially said that wherever resources lie, they must be available for sale to the highest bidder. In other words, whoever has the money to buy those resources has a legally defensible right to them. According to those rules, the oil of Venezuela belongs to the US every bit as much as if it lay under the soil of Texas or Missouri. Meanwhile technology, or "intellectual property," was regarded as proprietary; thus nations with prior investments in this strategy were at an advantage while "underdeveloped" nations were systematically discouraged from adopting it.

In the early 21st century, growing opposition to globalization — peaceful and otherwise — began to emerge in mass public demonstrations as well as in terrorist attacks. Most Americans, however, informed only by commercial media outlets owned by corporations with energy-resource interests, remained utterly in the dark as to what globalization was really about and why anyone would object to it.

In this first chapter, we have focused on energy principles in physics, chemistry, ecology, and sociology. We have noted how important energy is for the functioning of ecosystems and societies, and have traced its role in the history of the US rise to global dominance.

As we have just seen, America became the preeminent world power in the 20th century not just because of its professed ideals of freedom and democracy, its ingenuity, and the hard work of its people, but more importantly because of its immense wealth of natural energy resources and its ability to exploit them. For the past three decades, the depletion of those resources has been propelling US economic, political, and military policy in a certain definable direction, which we will explore further in Chapter 5.

In order to better understand these developments and their likely consequences, we need to examine more thoroughly the recent history of energy resources and their impact on societies around the globe. It is to this subject that we turn next.

2

Party Time: The Historic Interval of Cheap, Abundant Energy

In 1859 the human race discovered a huge treasure chest in its basement. This was oil and gas, a fantastically cheap and easily available source of energy. We did, or at least some of us did, what anybody does who discovers a treasure in the basement — live it up, and we have been spending this treasure with great enjoyment.

— Kenneth E. Boulding (1978)

Oil has literally made foreign and security policy for decades. Just since the turn of this century, it has provoked the division of the Middle East after World War I; aroused Germany and Japan to extend their tentacles beyond their borders; the Arab Oil Embargo; Iran versus Iraq; the Gulf War. This is all clear.

— Bill Richardson, Secretary of Energy (1999)

Whether we are talking of an individual citizen or a whole community, "cataclysmic wealth" can have disastrous consequences .... Its use rises sharply to create new habits and expectations. These habits are accompanied by an irrational lack of care about usefulness or waste. The process develops habits in individual people, and institutions in whole societies, which accustom them to operating on the basis of excess and wastefulness; and, although different episodes have different endings, one prospect sees the affected groups, long after the cloudburst of wealth has passed, trying every kind of expedient — borrowing, sponging, speculating — to try to ensure that the private habits or public institutions of excess and waste are maintained. The result is at best a measure of social disintegration; at worst, collapse.

— Barbara Ward (1977)

Forests to precede civilizations, deserts to follow.

— François René Chateaubriand (ca. 1840)

Fossil fuels have provided us with a source of energy so abundant and cheap that, in our rush to take advantage of them, we have utterly transformed our societies and our personal lives. This transformation has been so profound as to compare with the agricultural revolution of ten thousand years ago. However, that earlier development was, by comparison, an event in slow motion, requiring centuries to unfold in the areas where it originated, and millennia to reach most other inhabited regions of the planet. By contrast, fossil-fueled industrialism has swept the world in a mere two hundred years.

Historians are accustomed to speaking of the "Old Stone Age," the "New Stone Age," the "Bronze Age," or the "Iron Age" as a way of denoting certain periods by their characteristic technological regimes. An "age" in this sense may last tens of thousands of years, as did the Old Stone Age, or, in the case of the Bronze Age, only a millennium or so. The period of time during which humans will have discovered petroleum, reshaped their societies to make use of it, and then exhausted nature's supply promises to last little more than two centuries in total. This period of overwhelming transformative change has sometimes been called the "Petroleum Era" or the "Industrial Age," but, in view of its relative brevity, it may be more appropriate to call it the "Petroleum Interval" or the "Industrial Bubble."

This recent fossil-fuel-based explosion of human population and invention, though in many ways unprecedented in history, shares some basic characteristics with previous socio-technic transformations. Most great socio-technic revolutions begin out of necessity. When circumstances are comfortable, people tend to prefer doing things in old, familiar ways. It is when things aren't going well — that is, during times of an energy deficit, in any of its multitude of forms — that humans are most willing to experiment. But, having solved their immediate problems through some technical or social innovation, people often find that their new strategy has liberated more energy than was actually needed. Then, in developing ways to fully implement the new strategy and to take advantage of a sudden and unexpected energy abundance, they reshape their society, which typically grows in size and complexity.

The agricultural revolution illustrates this principle. Much evidence suggests that humans took up horticulture and then agriculture at least partly out of necessity; as anthropologist Marvin Harris has put it, "it seems clear that the extinction of the Pleistocene megafauna triggered the shift to an agricultural mode of production in both the Old and New Worlds."[1] But agriculture did not merely make up for the caloric loss resulting from hunting large prey animals to extinction (in fact, it did this only poorly); rather, it opened up an entirely new way of life — one that would eventually both enable far more humans to survive in closer proximity to each other than ever before and encourage the building of permanent and expanding settlements in which division of labor and class distinctions would emerge and proliferate.

As we are about to see, this same principle has been powerfully at work throughout the duration of the Industrial Bubble. Necessity led to invention, which led to growth and transformation.

In this chapter, we will trace the history of this fateful period from its beginnings to the present.

Energy in Medieval Europe

If we could somehow carry ourselves back in time to central and western Europe in the year 400 AD and fly a few hundred feet above that continent, our bird's-eye view would reveal a land covered from horizon to horizon by dense forest, with only occasional clearings. In each of those clearings we might see a cluster of thatched huts, with smoke rising from one or more wood fires.

The Europeans of 400 AD relied on an energy regime based mostly on wood. They built their houses and furniture with wood; they made tools from it, including plows, pumps, spinning wheels, and wine-presses; they made transportation devices (carts and boats) from it; and they used it as fuel with which to heat their homes and cook. Whatever bits of metal they used — blades, coins, jewelry, horseshoes, nails — came from wood or charcoal-fired hearths.

If wood was supremely useful, it was also abundant. A vast forest lay within sight of virtually every town or village. In addition to its immediate benefit of supplying fuel, the temperate oak forest of Europe also supported a profusion of wild game animals, including deer, boar, and numerous bird species, such as pheasant and quail. Human settlements were small, seldom numbering more than a few hundred people; the total population of Europe probably — exact figures are not known — did not exceed 25 million (compared to 600 million today, if European Russia is included).

That the ancient Europeans revered the forest is evidenced by their traditions concerning the sacredness of certain groves, by their customs of making sacrifices and offerings to trees, and by their extensive lore regarding tree-spirits.

But, with the coming of Christianity, these early pagan attitudes (the Latin *paganus* means "peasant") were gradually replaced by the idea that the wilderness is inherently fallen and corrupt, to be reclaimed only by pious human work. Far from fearing the overcutting of forests, later medieval Europeans saw the clearing of land as their Christian duty. Cutting the forest meant pushing back chaos, taming Nature, and making space for civilization.

While wood was the principal fuel in medieval Europe, it was far from being the only available energy source. Generally, civilized humans have two broad categories of energy needs: for lighting and heating on the one hand, and for motive power for agriculture and transportation on the other. Until recent times, these two categories of needs were usually served by two separate categories of energy sources.

Lighting and heating required fuel. In medieval Europe, the burning of wood (occasionally straw or dried animal dung was used) provided heating fuel for virtually everyone. Fuel for lighting came from the burning of wax, tallow, rushes, or olive oil — but was considered too costly for any but the wealthy, except on special occasions.

Motive power at first came primarily either from human labor or animal muscle, though these would later be supplemented by power from water and wind. Despite the fact that the human engine is capable of generating comparatively little power, much of the land in Europe — as well as in China — was tilled directly by humans using a hoe or spade, without the help of an animal-drawn plow. Because people typically eat less than draft animals do and because their efforts are intelligently directed, they often provide a more economical source of power than do oxen, horses, or mules.

In medieval Europe, as in the great civilizations of China, Rome, and the Near East, forced human labor was common. While in Germany and eastern England a substantial portion of the peasantry was made up of free persons who held and worked lands in common, most communities elsewhere came to be organized around manors controlled by lords whose right to land could be defended, when necessary, by full-time specialists in violence (soldiers, vassals, knights, and sheriffs). Agricultural tenants, in order to gain the right to cultivate a plot of land, were required to work a certain portion of each year on their landlord's estate. Serfs were bound to the land as quasi-slaves; and though they retained certain economic and legal rights, many existed perpetually on the verge of starvation. Ironically, however, it is also true that, in view of the many holidays and festivals celebrated in medieval societies, the typical serf back then actually enjoyed considerably more free time on a yearly basis than does today's typical American salaried worker.

In the early medieval period, most of the power for pulling plows and carts was provided by oxen. Only during the 12th century did horses come to be used as draft animals in any great numbers, this shift being due to the invention and widespread adoption of the horse-collar. Both before and after this time, horses were widely used for military purposes, a mounted cavalryman being both more mobile and more formidable than a footsoldier. In Spain and southern France, mules provided motive power for agriculture and transportation. Mules would later also become the primary source of animal power in regions of the Americas dominated by Spain — namely Mexico and most of South America. In addition to pulling plows and carts, oxen, horses, and mules also provided power for machinery: at first, for grain mills; later, for pumps to drain mines and for textile looms.

A significant implication of the use of large ruminant animals for traction was the necessity of growing food for them. Oxen, which can live on grass stubble and straw, were cheaper to maintain than horses, which also need grain. A horse typically requires between four and five acres of land for its food production; thus the use of traction animals reduced the human carrying capacity of the land while at the same time adding to it by enabling the plowing of larger fields. The net result varied. Animals were costly, and only a prosperous individual could afford to keep a horse. However, until the beginning of the 20th century, the trend was toward the increasing use of animal power. By 1900, Britain had a horse population of 3.5 million, consuming 4 million tons of oats and hay each year, thus necessitating the importation of grain for both animals and humans. In the US during the same period, the growing of horse feed required one quarter of the total available cropland (90 million acres).

Throughout the medieval period, human and animal power was increasingly supplemented by power from watermills and windmills. Watermills had been known from the time of ancient Greece; the Romans, Chinese, and Japanese employed them as well. The Romans had contributed the significant innovation of gears, which permitted the wheel to be moved to a vertical position and enabled the millstone to turn up to five times faster than the propelling wheel. Toward the latter days of their Empire, the Romans appear to have been taking increasing advantage of such equipment, perhaps because of a scarcity of slave labor, though such incipient industrial efforts subsided with the collapse of their civilization in the fifth century. However, in the 12th and 13th centuries, Europeans, led by the Cistercian monks, began using water wheels more extensively, and for a greater variety of purposes, than in any time or place previously. Windmills were costlier to operate than watermills, but could be built away from streams and could be used, for example, to drain water from the soil and to pour it into canals — hence the windmill's significant role in the reclamation of land in the Low Countries.

Originally, both windmills and watermills were primarily used for grinding grain, an otherwise arduous process. A first-century verse by Antipater of Thessalonica describes the perceived benefits of the water wheel in both mythic and human terms:

> Cease from grinding, ye women of the mill; sleep late even if the crowing cock announces the dawn. For Demeter has ordered the Nymphs to perform the work of your hands, and they, leaping down on top of the wheel, turn its axle, which with its revolving spokes, turns the heavy concave Nysirian millstones. We taste again the joys of primitive life, learning to feast on products of Demeter, without labor.[2]

Gradually, ingenious though anonymous inventors worked to develop and extend the use of windmills and watermills. One of the most important of these refinements consisted in the use of gears both to harness the machine's motive power to operate tools, such as saws and looms, and to operate several implements simultaneously. Eventually, mills would be used to pump water from mines, crush ores, make paper, and forge iron, among other tasks.

It should be noted that Europeans also harnessed wind power for transportation by means of sails. Sailing ships already had a long history throughout the Mediterranean as well as in China; during the medieval period their use gradually increased with improvements in shipbuilding and navigational technology, so that, by the latter part of the 16th century, European ships were conveying an estimated 600,000 tons of cargo annually. Many countries additionally maintained large fleets of sail-propelled warships.

The development of watermills and wind power in the Middle Ages could be said to have constituted the first industrial revolution. It was a period of sometimes explosive rates of invention and the development of Class B and C tools (including the printing press); but perhaps more importantly, it was the time when the very first Class D tools appeared, consisting of iron components for windmills and watermills, such as the heavy tilt-hammers used in iron forging.

Iron played no small part in this first industrial revolution. The use of iron can be traced back to the 15th century BC in the Caucasus, and cast iron and coal firing were known in China as early as the fifth century BC — developments not seen in Europe until the 14th century. Moreover, in China and India a high-quality carbonized steel (known in Europe as Damascus or damask steel) was being made as early as the second century — Europeans would not produce steel of equal quality until the 19th century.

However, despite being somewhat late on the scene with regard to such improvements, Europeans increasingly made use of iron during the medieval period, with demand for it often being stimulated by a long-simmering arms race. With crusades, wars, invasions, and peasant rebellions recurring throughout the period, there was constant need for more and better swords and pikes and — following the introduction of gunpowder (another Chinese invention) in the 14th century — for arquebuses, cannons, and iron bullets, all in addition to the cooking utensils, cauldrons, armor horseshoes, nails, and plowshares that were the day-to-day products of local smiths. Between the 11th and the 15th centuries, significant developments included the replacement of hand bellows by a hydraulic blowing machine and the invention of the blast furnace, permitting the production of cast iron and low-grade steel.

Demand for other metals — copper, bronze, gold, and silver — was also on the rise during this period. While the manors of the early medieval period were almost entirely self-sufficient, so that money was required only for the purchase of imported luxury goods, a gradually increasing trade required ever larger quantities of copper, silver, and gold coins.

The production of all these metal goods required fuel. Smelting necessitated high temperatures achievable only by the burning of charcoal, which is made by charring wood in a kiln from which air is excluded. The quantities of charcoal — and therefore of wood — that were required were far from negligible: the production of each ton of iron required roughly 1,000 tons of wood.

Altogether, the medieval energy economy — based on wood, water, and wind as well as on human and animal power — relied on resources that were renewable but not inexhaustible. Oak forests could regenerate themselves, though that took time. But trees were being cut faster than they could regrow, and the result was a rapid depletion of medieval Europe's primary fuel source.

While the construction of more and larger ships and the invention of the blast furnace contributed to the accelerated felling of trees, the ultimate cause was simply the increase in human population: many forests were cut merely to make way for more crops to feed people and domesticated animals. Prior to the Industrial Revolution of the late 18th century, there were two prolonged population surges in Europe: between 1100 and 1350 and between 1450 and 1650. Following the first surge, there was a sharp recession due to the Black Death; following the second, population growth tapered off partly due to recurring famines. From an ecological point of view, Europe had become saturated with humans, whose demands upon the environment were resulting in a rapid destruction of their temperate-forest ecosystem. Any further population growth would have to be based upon the acquisition of a new energy source.

Much of the southeast of England had been deforested by the end of the 11th century, and by 1200 most of the best soils of Europe had been cleared for agriculture. Wood shortages became commonplace in the 12th and 13th centuries. Between 400 and 1600 AD, the amount of forest cover in Europe was reduced from 95 percent to 20 percent. As scarcities appeared, wood began to be transported ever further distances by cart and by water. By the 18th century, blast furnaces were able to operate only one year in every two or three, or even only one in five or ten. Wood shortages led to higher prices for a variety of goods; according to Sully, in his *Oeconomies Royales*, "the price of all the commodities necessary for life would constantly increase and the growing scarcity of firewood would be the cause."[3]

In sum, the medieval period in Europe was a time of technological innovation, population growth, and energy-resource depletion within a region that, compared with China and the Islamic world, must be considered a cultural backwater. But this was the cultural, demographic, and geographic crucible for two immense developments. The first, whose significance was almost immediately recognized, was the commencement of the European age of exploration and colonization, which would eventually transfer vast wealth from the New World to the Old. The second was the gradually increasing use of a new kind of fuel.

The Coal Revolution

According to the report of an early missionary to China, coal was already being burned there for heating and cooking, and had been so employed for up to 4000 years.[4] Likewise in early medieval Europe, the existence of coal was no secret, but the "black stone" was regarded as an inferior fuel because it produced so much soot and smoke. Also, it occurred only in certain regions and had to be mined and transported. Thus, until the 13th century, it was largely ignored in favor of wood.

As wood shortages first began to appear, poor people began heating their homes by burning coal — most of which came from shallow seams and was a soft and sulfrous type that produced an irritating, choking smoke. Much was "sea-coal," which consisted of lumps collected from beaches and derived from cliff outcrops. By the late 13th century, London — a town of a few thousand inhabitants — was already cloaked in smog during the winter months. By the 16th and 17th centuries, even the rich were forced to make do with this inferior fuel. In the words of Edmund Howes, writing in 1631, "the inhabitants in general are constrained to make their fires of sea-coal or pit-coal, even in the chambers of honourable personages."[5]

However, coal was soon found to have advantages for some purposes — especially for metal working, since the higher temperatures possible with coal-fed fires facilitated the smelting of iron and other ores. Moreover, experimenters soon discovered that the roasting process used to make charcoal could be adapted to coal, the result being an extremely hot-burning fuel called coke. The use of coke in iron and steel production, beginning in England in the early 17th century, would so transform those industries as to constitute one of the key developments paving the way for the Industrial Revolution.

By the 17th century coal had revolutionized far more than metallurgy and home heating: its use had become essential for manufacturing glass, bricks, tiles, and salt (through the evaporation of sea water) as well as for refining sugar, brewing beer, and baking bread.

Meanwhile, the extraction of coal — a dreary, dangerous, and environmentally destructive activity at best — led by necessity to a series of important mechanical inventions, including the mechanical lift and the underground tunnel with artificial lighting and ventilation. As mines were sunk ever deeper, sometimes to a depth of 200 feet or more, water tended to accumulate in the bottoms of the shafts. Workmen drained the water either with hand pumps or bucket brigades. In 1698, Thomas Savery devised a pumping engine that condensed steam to create a vacuum to suck water from mineshafts. The engine was extremely inefficient, requiring enormous amounts of energy to lift modest quantities of water. Just ten years later, Samuel Newcomen introduced a self-acting atmospheric engine operating on different principles; and though it constituted the first crude steam engine, it was used solely for pumping water from coal mines: at that time, no one apparently envisioned its possible employment in manufacturing and transportation.

In addition to water seepage, coal miners faced another problem: that of transporting the coal from the depths of mines to rivers or ports. Typically, balks of wood were thrown down to facilitate the movement of coal-bearing wagons. In 1767, Richard Reynolds constructed a trail of cast-iron rails, running from Coalbrookdale to the Severn, to hold the wagon wheels on track. Scores of similar tramways were constructed during the following two decades; in all cases, traction was supplied by horses.

Toward the end of the 18th century, inventors began toying with the idea of using the new steam engine (by now greatly improved through the efforts of James Watt) for locomotive power. After the expiration of Watt's patent, a Cornish engineer named Richard Trevithick devised a new high-pressure engine and, in 1803, installed it on a carriage in which he made several journeys through the streets of London. But public highways were too rough to accommodate the steam carriage, and so the idea languished for another two decades until George Stephenson hit upon the idea of putting the steam locomotive on rails like those used in the tramways of coal mines. When hired by a group of Quaker investors to construct a railway from Stockton to Darlington in 1821, Stephenson built the first steam railroad; and eight years later his locomotive, named the *Rocket*, won a competition on the newly constructed Liverpool and Manchester Railway, demonstrating once and for all the superiority of the new technology over horse-drawn rail carriages.

Until the mid-19th century, all ships had traveled by renewable human or wind power. Beginning in the 1840s, steam power began to be applied to shipping; by the 1860s, new developments, such as the steel high-pressure boiler and the steel hull, enabled a typical steamship to transport three times as much cargo from China to Europe as a typical sailing ship, and in half the time.

The effects of these innovations on the economic life of Europe were dramatic. Trade was facilitated, both within nations (between the countryside and the city as well as between cities) and among nations and continents. More trade meant the extraction of more ores and other resources. The steam engine also greatly accelerated the transformation of those resources by industrial processes, as inventors devised a variety of steam-powered machines — including powered looms, cotton gins, lathes, die presses, and printing presses — to supplement or replace human labor. ·

Coal also had important chemical by-products. Of these, the earliest to have a significant social impact was manufactured (or artificial) gas. The first gaslights appeared in England in the 1790s, when William Murdock, an engineer and inventor, lit his own factory and then a large cotton mill in Manchester. The first gas street lighting was installed in London in 1807. In the United States,

Baltimore was the first city to light its streets with gas, in 1816. Paris adopted gas street lighting in 1820. Soon nearly every town with a population of over 10,000 had a gas works, and the discharge of coal tars from the production of manufactured gas was being blamed for drinking-water pollution and the contamination of crops. In 1877, inventor T. S. C. Lowe discovered a way to make "fuel gas" from steam enriched by light oils recovered from gas-making residual tars. Fuel gas (also known as "carburetted water gas") was seen by the gas industry as a means of combating the inroads being made by electricity on gas lighting.

With the discovery of coal-tar dyes in 1854, coal byproducts also gave rise to the establishment of the chemical industry. The new synthetic dyes revolutionized the textile industry and led to the growth of the German chemical and pharmaceutical companies Hoechst and I.G. Farben.

Coal was thus central to the pattern we call industrialism. Even wage labor seems to have originated in the mining industry, and, as Lewis Mumford once noted, the "eight-hour day and the twenty-four-hour triple shift had their beginning in [the coal mines of] Saxony."[6] By the late 19th century, the factory, with its powered machines, had revolutionized human labor, the economy, and society as a whole. First in England, and then in America, Germany, and a growing roster of other nations, settled cultivators and craftspersons became managers, wage-earning employees, or unemployed urban paupers; and economies that previously had been based on local production for local consumption became increasingly dependent on the long-distance trade of raw materials and finished goods.

One way of gauging the pace and extent of this transformation is to chart the quantities of coal being mined and used during the 19th century. In 1800, the annual world coal output stood at 15 million tons; by 1900, it had risen to 700 million tons per year — an increase of over 4,000 percent. In the last two years of the 19th century (1899–1900), the world used more coal than it had in the entire 18th century.

However, this vast expansion in coal usage was not evenly distributed over the globe. It occurred primarily in Europe and North America; and of all countries, Britain used by far the most — between one third and one half of the global total throughout the 19th century. (Germany, a rival industrial power, was also a significant user.) The ability of British industry to take advantage of this new energy resource had important geopolitical consequences: British cargo steamers carried raw materials from around the world to British ports, whence they were taken by trains to factories; manufactured goods were then hauled by train from factories to ports, whence they were distributed to colonies thousands of miles away. This system of trade was based both on policies, laws, and treaties that greatly favored the colonizing nation over the colonies, and on a highly mobile, industrialized form of military power capable of enforcing those laws and treaties. While colonialism had existed prior to the widespread use of fossil fuels, Britain's industrial version of it greatly intensified its essential practices of extracting wealth and consolidating political power.

Other nations envied Britain's colonial empire but lacked either the energy resources or the geographical prerequisites (such as coastlines and ports) or other historical advantages (including prior colonies and investments in industry).

America's energy path during the early decades of the 19th century differed greatly from that of Britain. Because of its abundant forests and numerous rivers, the United States relied primarily on wood- and water-power for its early industrial development; and during the first half of the century, much of the energy for agricultural production came from African slaves. As late as 1850, half of the iron produced in the US was smelted with charcoal. Locomotives and riverboats continued to burn wood well into the last two decades of the century, when America's forests began to be dramatically depleted. Only in the mid-1880s did a shift to coal begin in earnest; by 1910, coal accounted for three-quarters of the nation's energy supply. Like Britain, the US was favored with abundant indigenous deposits, especially in the mountainous regions of Pennsylvania, West Virginia, Kentucky, and Tennessee.

Global dependence on coal peaked in the early 20th century, when its contribution to the total world energy budget surpassed ninety percent. In the course of a hundred years, coal had transformed much of the world. New forms of production, new inventions and discoveries, new patterns of work, and a new geopolitical balance of power among nations were all due to coal. Cities were lit, and factory-made goods were produced in abundance. In addition, a great surge of population growth began, which was to be by far the most dramatic in world history.

The 13th-century Europeans who had reluctantly begun burning coal to heat their homes would scarcely have understood the ultimate implications of their actions. For them, coal was a sooty black stone with only a few practical uses. Surely, few stopped to think that, while all of their other energy resources were renewable (if exhaustible), coal was both exhaustible and nonrenewable.

At the population levels and scales of usage prevailing in the Middle Ages, the limits to coal must hardly have seemed imaginable. Nevertheless, a threshold had been crossed: from then on, an increasing proportion of the world's energy budget would be derived from a source that could not be regrown or reproduced on a timescale meaningful to humans.

The Petroleum Miracle, Part I

In the late 19th and early 20th century, another new source of energy began to come into use: petroleum. As had been the case with coal, few people at first had any inkling of the consequences of the increasing exploitation of this new energy resource. But as coal had so dramatically shaped the economic, political, and social contours of the 19th century, petroleum would shape those of the 20th.

Again, necessity was the mother of invention. As motorized machines proliferated during the 19th century, vegetable oils, whale oil, and animal tallow were typically used for machine lubrication, and whale oil as fuel for lamps. Toward the end of the century, commercial whale species were being hunted to the point of extinction, whale oil was becoming increasingly costly, and tallow and vegetable oils were proving inadequate as lubricants for the ever-larger and more sophisticated machines being designed and built.

Petroleum had been known for centuries, perhaps millennia, and had been used in warfare as early as 670 AD, when Emperor Constantine IV attached flame-throwing siphon devices to the prows of his ships to spew burning petroleum on enemy vessels. Oil also had a long history of use for sealing and lubrication, and even for medicinal purposes. However, the exploitation of petroleum was limited to small quantities that seeped to the ground surface in only a few places in the world.

Beginning with the successful drilling of the first commercial oil well by "colonel" Edwin L. Drake in northwest Pennsylvania in 1859, petroleum became more widely available as a cheap and superior lubricant and, when refined into kerosene, as lamp fuel. The problems of whale-oil depletion and machine lubrication had been solved. But, of course, oil soon would prove to be useful for many other purposes as well.

Fortunes were quickly made and lost by dozens of drillers and refiners, as a rapidly expanding supply of petroleum fed the nascent demand. By 1866, Drake himself was bankrupt; meanwhile, an extraordinarily business-savvy early oilman named John D. Rockefeller had begun purchasing crude in Pennsylvania, Ohio, and West Virginia and refining it under the name Standard

Oil. Soon he had the largest refining operation in the country and was absorb-ing his competitors, using selective price-cutting strategies and obtaining kickbacks from the railroads that transported both his and his competitors' crude.

Rather than buying up the other refiners and producers outright, Rockefeller set up a trust by which stockholders in Standard Oil controlled the stock in dozens of other oil companies as well. Rockefeller's business strategy was sim-ple and consistent: be the low-cost producer, offer a reliable product, and ruthlessly undercut and assimilate any competitor. In addition to its refineries, Standard developed its own production and distribution systems, building pipelines and the first oil tankers. By 1880, Standard controlled ninety percent of the oil business in the US — and the rest of the world as well.

Rockefeller used Standard's domestic business tactics of predatory pricing, secrecy, and industrial espionage to absorb foreign oil companies — especially those in Europe, where industrialization and urbanization were stimulating an ever-increasing demand for kerosene and lubricating oil. Kerosene quickly became the foremost US-manufactured export; and Standard, with its European subsidiaries, became perhaps the first modern transnational corporation. In a mere decade and a half since founding Standard in 1865, Rockefeller had nearly achieved the goal he envisioned from the start: a worldwide monopoly on petroleum.

However, that monopoly hinged at least in part on the control of global production, a control that was soon threatened by the discovery of major reserves outside the American northeast. The first such threat emerged from the Russian empire, where oil was discovered in 1871 in Baku, a region on the Aspheron Penninsula in the Caspian Sea. Ludwig Nobel, known as the "Russian Rockefeller" — and brother of Alfred Nobel, the discoverer of dynamite and donor of the Nobel Prize — arrived in Baku at the beginning of the oil rush there and quickly established commercial dominance in production and refin-ing. By 1885, Russian crude production was at about one-third of the American production levels. But since demand within Russia itself could not absorb such an amount, the Nobels sought foreign markets. Help came from the French branch of the Rothschild banking family, which, over the previous century, had financed wars, governments, and industries, and now owned a refinery at Fiume on the Adriatic. The Rothschilds bankrolled a railroad from Baku to Batum, a port on the Black Sea, enabling the Nobels' oil to flow to European markets.

The Rothschilds soon bought their own oil wells and refineries in Baku, entering into competition with the Nobels. They also expanded the distribution of Russian oil to Britain, prompting Standard to set up its own affiliate in London, the Anglo-American Oil Company. The Rothschilds then looked further afield, namely to Asia, seeking still more markets for the ever-growing supply of Baku crude. In the early 1890s, they contracted with international trader Marcus Samuel to build a system of distribution throughout South and East Asia. Samuel began by embarking on an Asian tour; soon he was supervising the construction of storage tanks throughout Asia, undertaking major improvements in tanker design, and obtaining the right of passage through the Suez canal, which had previously been denied to oil shipping. Samuel's objective was nothing less than to beat Standard Oil at its own game, offering exported Russian oil throughout the Far East at prices Rockefeller could not match. Samuel's company — at first called the M. Samuel Company, later Shell Transport and Trading — achieved a coup that could hardly escape Standard's notice.

During the 1890s, Rockefeller, the Nobels, the Rothschilds, and Samuel engaged in what became known as the Oil Wars. Periods of price-cutting were punctuated with attempts at takeovers or grand alliances.

At the same time, oil production in the Dutch East Indies (now Indonesia) was growing at a furious pace under the commercial control of the Royal Dutch Company, which offered still another challenge to Standard's international dominance.

Further, while all of this global competition was intensifying, the oil business was changing in fundamental ways. With Thomas Edison's promotion of electric lighting in the 1880s, demand for kerosene peaked and began to recede. However, new uses for petroleum more than took up the slack. Oil-burning furnaces appeared toward the end of the century, as well as oil boilers for factories, trains, and ships — all promoted by Standard. By 1909, half of all petroleum extracted was being sold as fuel oil. But by far the most important new use of petroleum was as fuel for the internal combustion engine, developed in the 1870s by German engineer Nikolaus Otto. Gasoline, when first discovered, had, because of its extreme volatility, been regarded as a dangerous refinery waste product; when used in lamps, it caused explosions. Initially, it was simply discarded or sold for three or four cents per gallon as a solvent. Now it was seen as the ideal fuel for the new explosion-driven internal combustion engine.

Another important development was the appearance of a market for natural gas. The latter had frequently been found together with oil (most oil fields have gas deposits). Gas was also often found in coal deposits, and many early coal miners had died from asphyxiation from deadly "coal gas" or from gas explosions. Natural gas is mostly methane, but it also contains small amounts of ethane and heavier hydrocarbon gases, such as butane, propane, and pentane. During the first few decades of oil drilling, natural gas was often regarded as having no value and was simply flared (burned off). However, as prices for manufactured gas for street lighting rose and as environmental hazards from its production became more apparent, natural gas was seen as a cheap and environmentally more benign substitute. In 1883, Pittsburgh became the first city to replace manufactured gas with the cheaper natural gas. Three years later, Standard Oil formed the Standard Natural Gas Trust. But within a few years, electric street lighting and home lighting appeared as commercially viable options. As gas lights were gradually replaced with electric lights, local gas works were sold and consolidated, their infrastructure of pipes converted to the distribution of natural gas for cooking and heating.

Even though petroleum production, refining, and distribution had become huge and quickly growing commercial enterprises, the 19th century was the century of coal to its very end: only after the turn of the 20th would the world witness the true dawning of the petroleum era.

Electrifying the World

Before we continue with the story of petroleum, it is necessary to survey another energy development that would shape the 20th century in nearly as profound a way as would oil: electrification. Unlike petroleum or coal, electricity is not a source of energy, but rather a carrier of energy, a means by which energy can conveniently be transmitted and used. Electrification enabled the development and wide diffusion of home conveniences, business machines, and communication and entertainment devices — all connected by miles of wire to a variety of energy sources ranging from coal or oil boilers to hydroturbines to nuclear fission reactors. By making energy easy to access and use, electricity stimulated the use of energy for ever more tasks until, by the 20th century's end, most people in industrialized cities were spending virtually every moment of a typical day using one or another electrically powered device.

The first electric generator was invented in London in 1834, but decades elapsed before electricity saw commercial applications. Thomas Edison (1847–

1931), a former railroad telegrapher, began his career as an inventor by devising improvements to the telegraph and telephone. His laboratory was described as an "invention factory": Edison was the first to apply industrial methods to the process of invention, hiring teams of engineers to work systematically to devise new commercial technologies. This was a strategy widely adopted throughout corporate America in the following century. In 1878 Edison turned his attention to the electric light; in 1879 he lit his factory with electricity; and three years later, his workers installed carbon-filament electric lamps in the financial district of lower Manhattan. Edison, ever the astute businessman, supplied the entire system of generators, transmission lines, and lights, taking care to price electricity at exactly the equivalent of the price of piped gas.

Edison's system for generating and distributing electric power used current flowing in one direction only — direct current, or DC. This had the disadvantage of requiring neighborhood generating stations, since direct current was rapidly dissipated by resistance along transmission lines, given the technology then available (today high-voltage direct current can be transmitted long distances with relatively little line loss). Indeed, most of the factories and homes lit by early DC systems maintained dynamos on-site — which were both expensive and annoying: the generator in the basement of J. P. Morgan's mansion in New York made so much noise that his neighbors frequently complained. At that time, virtually no one envisioned the regional, centralized electric distribution systems we now take for granted.

Engineers of the time knew that alternating current, or AC — where electricity flows back and forth along transmitting wires, alternating its direction many times per second — was a theoretical possibility and could overcome the transmission limitations of direct current. However, no one had yet solved the basic technical problems, and no practical AC motor yet existed.

In 1884, a Serbian-American inventor named Nikola Tesla (1854–1943) approached Edison with his designs for an AC induction motor. Though Edison recognized the younger man's exceptional intelligence and immediately hired him to improve existing DC dynamos, Tesla's plans were ignored. The two men were utterly dissimilar in their approaches to invention: while Edison was a tinkerer with little understanding of theoretical principles, Tesla was a supreme theoretician comfortable with advanced mathematics. Tesla would later write:

> If Edison had a needle to find in a haystack, he would proceed at
> once with the diligence of the bee to examine straw after straw until

he found the object of his search. I was the sorry witness of such doings, knowing that a little theory and calculation would have saved him ninety percent of his labor.[7]

Tesla broke with Edison after a financial dispute (the latter offered a $50,000 bonus for a job he thought impossible, then refused to pay when Tesla accomplished the task), obtained financing, equipped his own laboratory, and proceeded to design, build, and patent the first AC motors. A Pittsburgh industrialist named George Westinghouse heard of Tesla's work, purchased rights to his patents, and went into competition with Edison.

By 1893, the financier J. P. Morgan had engineered a takeover of Edison's company and several other electrical device manufacturers; the resulting General Electric Corporation then settled into a "War of the Currents" with the Westinghouse company over the future of electrification. GE publicists made absurd claims about the dangers of AC power, while spectators at the Columbian Exposition were awed by the most impressive demonstration of electric lighting yet seen —provided under contract by Westinghouse. Flags fluttered, a chorus sang Händel's *Hallelujah Chorus*, and electric fountains shot jets of water high into the air. Twenty-seven million people attended the fair during the following months, all witnessing the practical wonders of alternating current. Tesla's victory was further underscored when the Niagara Falls Power Project, completed in 1896, chose Tesla's advanced AC polyphase designs for its giant dynamos. J. P. Morgan would later comment that his backing of Edison's DC system had constituted the single worst business decision of his career.

Tesla went on to invent or provide the theoretical foundations for radio (while Marconi is still usually given credit for this invention, the US Supreme Court affirmed the priority of Tesla's patents in 1943), robotics, digital gates, and even particle-beam weapons. While other scientists gleaned much of the credit for many of these developments and corporations made fortunes from them, Tesla preferred the role of lone visionary, giving yearly press interviews in which he articulated colorfully his plans for worldwide broadcast power, communication with other planets, and cosmic-ray motors. Tesla died nearly penniless at the height of World War II; his papers were immediately seized and sequestered by the Office of Alien Property. Though mostly forgotten for decades, Tesla is now widely regarded as the true father of 20th-century electrical technology.

At the turn of the century, as the War of the Currents was cooling off, the Utilities Wars were just heating up. Cities wanted electric street and home lighting, but controversy raged over the question of whether the utilities responsible for delivering the electricity should be privately or publicly owned. Financiers like J. P. Morgan and Samuel Insull (of Chicago's Commonwealth Edison) lobbied to make utility monopolies a perpetual "dividend machine" for investors while public-power advocates in hundreds of towns and cities around the country insisted that costs to consumers should be controlled through public ownership. In most cases, the private interests won, resulting in the creation of giant utility corporations like Pacific Gas & Electric (PG&E) and Continental Edison; however, scores of communities succeeded in creating publicly-owned municipal power districts that typically sold electricity to consumers at much lower prices.

During the 1930s, President Franklin D. Roosevelt battled the private utility interests in his campaign for rural electrification. His largest and most successful public works project, the Tennessee Valley Authority, at first administered by the visionary Arthur Morgan, built dams and distribution systems, making electric power almost universally available throughout the rural East. Today, rural power co-ops and municipal power authorities still control about 20 percent of electrical generation and distribution in the country, and recent developments, such as the bankruptcies of Enron and PG&E — with executives profiting handsomely while customers and employees paid dearly — have revived the public-power movement throughout the nation.

Even before the beginning of the 20th century, electricity already had many uses in addition to lighting. Electric streetcars and subways began replacing horse-drawn streetcars in the larger cities in the 1890s, and this led to a major change in urban development patterns: the growth of suburbs. In 1850, the edge of the city of Boston lay a mere two miles from the city center; by 1900, electrified mass transit had allowed the city perimeter to spread ten miles from the business district. Previously, city centers had been the most densely populated areas; now, urban cores began emptying as residents moved to the suburbs, leaving the heart of town to financial and commercial activity.

As factories were electrified, opportunities for automation cascaded, further fragmenting production tasks and eliminating the need for skilled labor. A former Edison employee named Henry Ford recognized these possibilities early on, and the electrified assembly line he created for the production of his motorcars stood as an example for other manufacturers of how to cut costs and ensure uniform quality. A Model T Ford sold for $825 in 1908, but by 1916, automated mass production had brought the price down to $345.

As homes were electrified, even domestic work began to be automated, and housewives were bombarded by advertisements informing them of the potential gains in "productivity" and "efficiency" available through the use of gadgets ranging from vacuum cleaners and washing machines to electric toasters, mixers, and irons.

Because of its unique properties, electricity was used in ways that would not have been possible with other forms of energy. The field of electronics — encompassing radio, television, computers, and scores of other devices based first on vacuum tubes and later on semiconductors — would revolutionize communications and entertainment as well as information storage and processing.

While electricity offers extreme convenience to the user, it is an inherently inefficient energy carrier. For example, when coal is burned to drive dynamos, only 35 percent of its energy ultimately becomes electricity. Inefficiencies are also inherent in transmission lines and in end-use motors, lights, and other powered devices. However, as long as primary energy sources remain cheap, such inefficiencies can be easily afforded. At current prices, an amount of electricity equivalent to the energy expended by a person who works all day, thereby burning 1,000 calories worth of food, can be bought for less than 25 cents.

Coal is still the principal primary energy source for the generation of electricity in the US and throughout the world. In 2004, public and private US electric utilities derived 51 percent of their power from coal, 20 percent from nuclear fission, 7 percent from hydro, 16 percent from natural gas, 3 percent from oil, less than 1 percent from wind and photovoltaics, with the remainder coming from other alternative sources.

Electricity's availability for a vast range of tasks has led to a massive increase in total energy usage. Whole industries — such as aluminum production — have arisen that are completely dependent upon electricity. On the whole, during the 20th century electricity consumption increased twice as fast as the overall energy consumption.

The Petroleum Miracle, Part II

In the first half of the 20th century, as electricity was revolutionizing homes and workplaces, the industrialized world's reliance on coal gradually subsided while its use of oil expanded greatly, reshaping nearly all spheres of life.

The structure of the petroleum industry underwent significant shifts in the early years of the century. In 1902, Samuel was forced to merge his company with Royal Dutch to create Royal Dutch-Shell. And with the discovery of oil in Persia (now Iran), the Anglo-Persian Oil Company (later British Petroleum, or BP) came into being. Persian crude would be the first commercial oil to come from the Middle East.

Meanwhile, in the US, Rockefeller's cutthroat business tactics and near-monopolization of the domestic industry led to an anti-trust suit brought by the Federal government. In 1911, a decision of the Supreme Court forced the breakup of Standard Oil Company into Standard Oil of New Jersey (which later became Exxon), Standard Oil of New York (Mobil), Standard Oil of California (Chevron), Standard Oil of Ohio (Sohio, later acquired by BP), Standard Oil of Indiana (Amoco, now BP), Continental Oil (Conoco), and Atlantic (later Atlantic Richfield, then ARCO, then Sun, now BP). Rockefeller eventually profited handsomely from the split, and the new companies carefully avoided directly competing with one another.

At the turn of the century, Russia had briefly become the world's largest oil producer. However, political upheavals in that country undermined the further development of the industry. Soon new discoveries in the US — in California, Texas, and Oklahoma — made America again the foremost oil-producing and -exporting nation, a position it would hold for the next half century. In 1901, the Spindletop oil gusher in Texas marked a shift in the center of gravity of American production away from the Northeast and toward the Southwest. Further spectacular Texas and Oklahoma discoveries in the 1930s led to dramatic overproduction and price volatility: for a short time, the price of crude fell to four cents per barrel, making it literally cheaper than drinking water. The Texas and Oklahoma discoveries also engendered several new companies, including Texaco and Gulf Oil.

These developments taken together resulted in the domination of the world petroleum industry throughout the rest of the century by the so-called "Seven Sisters": Exxon, Chevron, Mobil, Gulf, Texaco, BP, and Shell. By 1949, the Seven Sisters owned four-fifths of the known reserves outside of the US and the USSR and controlled nine-tenths of the production, three-quarters of the refining capacity, two-thirds of the oil-tanker fleet, and virtually all of the pipelines.

Throughout the first decades of the century, the US was in a position to control the world oil price. This changed in the second half of the 20th century, as we will see shortly, when US production declined while production in the Middle East increased.

During the first half of the century, as electricity steadily eroded the market for kerosene as a source of illumination, new uses for petroleum products stoked ever greater demand for oil. By 1930, gasoline was the principal refined product of the petroleum industry, and aviation fuel was beginning to account for a noticeable share of oil production. And as the chemical industry switched from coal tar to petroleum as raw material, new synthetic materials — nylon and a wide range of plastics — began to replace traditional materials such as wood, metal, and cotton in manufactured consumer products.

It is difficult to overstate the extent of the transformations of the world economy, of industry, and of daily life that can be attributed to the use of petroleum during the 20th century. We can perhaps appreciate these transformations best if we discuss separately the fields of agriculture, transportation, and warfare.

Agriculture

One of the greatest problems for agriculture had always been the tendency of soils to become deficient in nitrogen. The traditional solutions were to plant legumes or to spread animal manures on the soil. But these nitrogen sources were not always adequate. In 1850, explorers of islands off the coast of Chile and Peru had discovered entire cliffs of guano — the nitrogen-rich excreta of sea birds. Over the next two decades, 20 million tons of guano were mined from the Chilean and Peruvian islands and shipped to farms in Europe and North America. Once that supply was exhausted, the search was on for a new source of usable nitrogen.

In 1909, German chemists Fritz Haber and Carl Bosch devised a method for fixing atmospheric nitrogen by combining it with hydrogen to make ammonia. At first, the process used coal to fuel the machinery and as a source of hydrogen; later, coal was replaced by natural gas. As geographer Vaclav Smil has argued in *Enriching the Earth: Fritz Haber, Carl Bosch, and the Transformation of World Food Production*[8], the Haber-Bosch process probably deserves to be considered the principal invention of the 20th century since today ammonia synthesis provides more than 99 percent of all inorganic nitrogen inputs to farms — an amount that roughly equals the nitrogen tonnage that all of green nature gains each year from natural sources (legumes, lightning strokes, and animal excreta). More than anything else, it is this doubling of available nitrogen in the biosphere that has resulted in a dramatic increase in food production throughout the century, enabling in turn an equally dramatic increase in human population. At the same time, however, the widespread agricultural application of synthetic ammonia has led to nitrogen runoffs into streams and rivers — one of the most significant pollution problems of the last century.

Agriculture was also revolutionized by tractors and other motorized equipment as well as by motorized systems of distribution. Previously, one-quarter to one-third of all agricultural land in North America and Europe had been devoted to producing feed for the animals that pulled plows and wagons; thus the replacement of animals by motorized equipment meant that more land could be freed for human food production. Also, because tractors could cover more ground more quickly than draft animals, fewer farmers were needed to produce an equivalent amount of food; hence larger farms became economically feasible, indeed advantageous. As small subsistence farms were increasingly put at disadvantage, more farmers left the countryside to seek work in the cities.

The development of petrochemical-based herbicides and pesticides after World War II increased yields even further. In the 1960s and '70s, international development agencies promoted the use of motorized farm equipment, synthetic ammonia fertilizers, and chemical herbicides and pesticides throughout the less industrialized nations of the world; known as the "Green Revolution," this program resulted in predictably enhanced yields, but at horrendous environmental and social costs.

Transportation

The transportation revolution of the 20th century had social, economic, and environmental consequences that were nearly as profound as those in agriculture. Central to that revolution were the automobile and the airplane — two inventions dependent on the concentrated energy of fossil fuels.

In its early days, the automobile — invented in 1882 by Carl Benz — was a mere curiosity, a plaything for the wealthy. Nevertheless, the idea of owning a private automobile was widely and irresistibly attractive: the young and the upwardly mobile could not help but be seduced by the motorcar's promise of speed and convenience — even though the reality of journeying any distance in one actually entailed considerable inconvenience in the forms of noise, dust, mud, or mechanical breakdown. Promoters of the automobile claimed that widespread car ownership might relieve the nuisance of horse feces covering urban streets (for cities like New York and Chicago, this posed a serious pollution dilemma), but few gave much thought to the problems that near-universal automobile ownership might itself eventually entail.

One of the principal hindrances to the growth of the early auto industry was the lack of good roads. In order to travel at the speeds of which they were capable, automobiles needed surfaces that were smoothly paved — but few existed. Demand for more public funding for highways was already growing, fed partly by bicyclists, but motorists added dramatically to the public pressure. Beginning at the turn of the century, car owners and manufacturers, together with oil and tire lobbyists, succeeded in persuading all levels of government to, in effect, subsidize the automotive industry — at a rate that would cumulatively amount to hundreds of billions of dollars — through appropriations for road construction.

Henry Ford, America's most prominent automotive industrialist, proposed to make cars so cheap that anyone could own one. Ford made sure to pay his factory workers enough so that they could afford to buy a coupe or sedan for themselves. However, as inexpensive as the Model T was by today's standards, it still represented a cash outlay beyond the means of most American families.

Automated, fuel-fed mass production was proving capable of turning out goods in such high quantity as to overwhelm the existing demand. Until this time, the average family owned few manufactured goods other than small items such as cutlery, plates, bowls, window glass, and hand tools. Virtually none had motorized machines, which were simply too expensive for the typical family budget. The industrialists' solutions to this problem were advertising and credit. More than any other product, the automobile led to the dramatic expansion, during the 1920s, of both the advertising industry and consumer debt. Car companies nearly tripled their advertising budgets during the decade; they also went into the financing business, making car loans ever easier to obtain. By 1927, three-quarters of all car purchases were made on credit, and there was one car for every 5.3 US residents.

That same year, 1927, was the first in which there were more people buying a car to replace a previous one than there were people buying a first one. As the roaring twenties drew to a close, the market for automobiles became saturated while American families saddled themselves with a record amount of consumer debt. But car companies kept producing more Fords, Buicks, Hupmobiles, and Stutzes, thus setting the stage for a recession. The auto industry was not solely responsible for the full-blown economic catastrophe that followed — overly lenient rules governing stock speculation played a prominent role as well — but it contributed in no small way to the ensuing bankruptcies, bank failures, and layoffs.

By now, the automobile manufacturers together controlled a significant proportion of the national economy: General Motors was the world's largest corporation, with Ford and Chrysler following closely behind. Whatever the Big Three automakers did sent ripples through the stock market, the banking system, and national labor organizations. The auto industry had united the interests of other giant industries — oil, steel, rubber, glass, and plastics — in the manufacturing, fueling, and marketing of a single product, and in the transformation of the American landscape, lifestyle, and dreamscape to suit that product. Subsidiary businesses sprang up everywhere — from spare parts distributors to local gas stations and repair shops, from fast-food chains to drive-in theaters.

Urban sprawl, which had begun in a few large towns with the installation of electric trolleys, exploded discrete cities into "metropolitan areas" with few clear boundaries, rolling on for mile after mile along major arteries. In New York, urban planner Robert Moses — who himself never drove — put the automobile at the center of his design priorities, creating grand new bridges and freeways for commuters while gutting entire neighborhoods to make way for on-ramps and off-ramps. For over forty years, from the 1930s to the late '70s, Moses rebuilt Manhattan to suit motorists; at the end of the process, traffic and parking problems were worse than they had been at the beginning, and the city had sacrificed much of its charm, neighborhood integrity, and historical interest along the way.

Many European cities responded to the automobile differently by investing more in trains, trolleys, and subways. Partly as a result, per capita auto ownership in Europe for a time remained significantly lower than in the US; meanwhile, the narrowness of old European city streets and the higher price of fuel encouraged the design of smaller cars.

The European approach to mass transit could have taken hold in the US, which maintained excellent inter-urban passenger rail lines and many fine urban streetcar systems until mid-century. However, in 1932 General Motors formed a company called United Cities Motor Transit (UMCT), which bought streetcar lines in town after town, dismantled them, and replaced them with motorized, diesel-burning buses. In 1936, GM, Firestone, and Standard Oil of California formed National City Lines, which expanded the UMCT operation, buying and dismantling the trolley systems in Los Angeles and other major cities.

By 1956, 45 cities had been relieved of their electric rail systems. The bus services that replaced them were, in many instances, poorly designed and run, leaving the private auto as the transportation mode of choice or necessity for the great majority of Americans. Public transportation in America reached its broadest per-capita usage in 1945, then fell by two-thirds in the succeeding twenty years.

The American love affair with the auto was also encouraged by what would become the biggest public-works project in history: the Interstate Highway System. Modeled on Hitler's Autobahn, the Interstate System came into being through the Interstate Highway Act, passed in 1956 partly as a measure for national defense. The bill authorized $25 billion for 38,000 miles of divided roads; by comparison, the entire national budget in 1956 was $71 billion, and the Marshall Plan had cost only $17 billion. It was the Interstates, more than anything else, that would eventually nearly destroy the American passenger rail system: the trains simply could not compete with so highly subsidized an alternative.

Car ownership meant convenience, power, and even romance, as the typical young couple found freedom and privacy in the back seat of the parents' Chevy. Soon they'd be married, the husband commuting to the office, his wife chauffeuring the kids to music lessons and little-league games. The gift or purchase of a first car would become as important a rite of passage for every teenager as graduation from high school. Life would become unimaginable without the Mustang, Camaro, or Barracuda in the driveway, ready and waiting for adventure.

However, the love affair with the car always had its dark side. While in Paris in 1900, novelist Booth Tarkington overheard and recorded the comment, "Within only two or three years, every one of you will have yielded to the horseless craze and be the boastful owner of a metal demon ... Restfulness will have entirely disappeared from your lives; the quiet of the world is ending forever."[9] But noise would prove perhaps the least of the car's noxious effects; air and water pollution, the loss of farmland due to road construction, and global warming constitute far worse damage. Car culture has also resulted in the disappearance of wildlands and poses a constant danger to animals: the toll in road kill is about a million wild animals per day in the US alone.

Out-of-pocket expenses for car ownership today average about $1,500 per vehicle per year. But if all of the environmental and social losses were factored in, that cost would be closer to $25,000 per car, according to some calculations.

One of the greatest of those "external" costs is car crashes: since 1900, more than twice as many Americans have died in auto collisions than have been killed in all of the wars in US history.

If the dollar cost of motoring is burdensome, the energy cost is staggering. The typical North American driver consumes her or his body weight in crude oil each week, and the automobile engines sold this year alone will have more total horsepower than all of the world's electrical power plants combined. Globally, cars outweigh humans four to one and consume about the same ratio more energy each day in the form of fuel than people do in food. A visitor from Mars might conclude that automobiles, not humans, are the dominant life form on planet Earth.

Like the automobile, the airplane began as an unreliable plaything, but one that evoked the promise of the superhuman power of flight. Its potential ability to speed up travel and to skip over geographic obstacles led to its early use in mail service. The first regularly scheduled passenger service began in the 1920s, though only the wealthy and the adventuresome took advantage of it; everyone else took the train or drove.

In the late 1950s, the first passenger jets entered service. At that time, most long-distance travelers still relied on cars, trains, buses, and ships; only the elite comprised the "jet set." But gradually, as tickets became affordable, more people began to board jet planes; by the 1970s, flying had become a standard mode of long-distance travel, especially for transoceanic journeys, and airports had taken on the former function of train or bus stations. This was especially true in the geographically far-flung cities of North America, where options for long-distance travel gradually narrowed to two: car or plane.

The growth of air transport vastly expanded the tourist industry, from 25 million tourists in 1950 to nearly half a billion by the year 2000. Hotels, travel agencies, and restaurants benefited enormously; today many cities and some nations are largely supported by jet-transport tourism.

Due to significant improvements in jet engine design in the past decades, the typical airline passenger experiences a miles-per-gallon efficiency roughly equivalent to — and, in some instances, better than — that of an automobile driver. Today, roughly ten percent of extracted oil is refined into kerosene to fuel jets. Americans now fly a total of 764 million trips per year — 2.85 airplane trips per person, averaging 814 miles per trip.

Aviation is the only transport form not significantly regulated to reduce environmental impact. Airports are typically sites of extreme air pollution, and jet aircraft contribute substantially to the destruction of the atmospheric ozone layer.

Warfare

At the beginning of the 20th century, wars were being fought with mounted cavalry, foot soldiers, horse-drawn artillery, and coal-fired warships. In the First World War, military strategists began to appreciate the advantages of applying more sophisticated fossil-fueled technology to the project of killing. Warships, and especially submarines, were converted to oil or diesel power, thus giving them a longer range and greater speed; and tanks and motorized troop carriers (of which the first were simply commandeered Parisian taxicabs) began to revolutionize ground warfare. Meanwhile, airplanes offered the possibility of improved reconnaissance and of raining terror from the skies.

The outcome of World War I was largely determined by oil: the Allies blockaded German supply routes, while Germany sought to cut off shipments to Britain with submarine warfare. The US, the world's largest petroleum producer, was a significant help to the Allies. When the Allies succeeded in denying Germany access to Romanian oil fields, German industry began to suffer from a shortage of fuels and lubricants. By 1917, civilian trains were no longer in service, and airplanes were running poorly on substitute fuels. On November 11, with its army in possession of only days' worth of essential fuels, Germany surrendered.

The lessons of this defeat were not lost on Adolf Hitler, who promised to reverse disastrous economic and social conditions resulting from the humiliating terms of the Versailles peace treaty. Germany could not fight again without adequate fuel stocks, nor could it allow itself to become bogged down in another war of attrition. Thus when Nazi generals began planning for the invasions that would precipitate World War II, they had two objectives in mind: access to oil supplies and swift, decisive victories through the use of surprise motorized attack — *blitzkrieg*. Among Hitler's principal objectives in Poland and the Soviet Union was control of the oil fields in those regions. When the Allies were eventually able to deny Germany access to those oil fields and to cut off German supply lines, the Nazi war machine simply ran out of gasoline.

In the Pacific, Japan — which had practically no indigenous oil resources — attacked Pearl Harbor after the US cut off oil exports in an effort to thwart Japanese imperial ambitions throughout the Far East. A major Japanese objective in the war was to secure oil fields in the Dutch East Indies. However,

American submarines succeeded in sinking enough tankers carrying oil from the East Indies to Japan that, by 1944, Japanese ships and planes were denied adequate fuel. By 1945, Japanese air pilots could no longer be given navigational training and Japanese aircraft carriers could no longer afford to take evasive action — all for lack of fuel.

Thus, by mid-century, oil had established itself both as an increasingly critical fuel for warfare and as an increasingly frequent geopolitical objective of war. Warfare had also become far more deadly, especially for noncombatants. All three trends would accelerate in the second half of the century.

Oil, Geopolitics, and the Global Economy: 1950–1980

At of the turn of the 20th century, Russia had begun the process of industrialization only in a few cities; throughout most of its vast territory, peasants worked their fields much as they had for centuries. Following the Bolshevik Revolution of 1917 and the subsequent period of turmoil and political reorganization, the leaders of the new Union of Soviet Socialist Republics decided to undertake a program of forced industrial development. Subsequently known as Stalinization, the program involved compelling agricultural workers into industry and consolidating peasant land holdings into giant collectives. Between 1927 and 1937, iron output increased by 400 percent, coal extraction by 350 percent, electric power generation by 700 percent, and the production of machine tools by 1,700 percent. However, the human consequences were horrific. Millions of peasants in the countryside died of starvation, and conditions for urban factory workers were abysmal. Political dissent of any kind was brutally crushed.

At the end of World War II, with most of Europe in ruins, the US and USSR emerged as victors. Their alliance against the Axis powers did not persist into peacetime as the two superpowers set about competitively dividing much of the world between them. Both had huge petroleum and coal reserves, but their histories and economic systems were fundamentally different. While never directly confronting one another militarily, the US and USSR waged proxy battles over resources and influence throughout the ensuing 45 years. The Soviet-dominated world was characterized by centralized, government-planned control and distribution whereas US-dominated nations were subsumed under the increasing power of giant multinational corporations.

The United States was the world's largest *consumer* of oil, its economy having been the first to widely exploit the use of petroleum through the mass production of automobiles and the development of a civilian airline industry. Meanwhile, as more Middle Eastern reserves were discovered and tapped, that area became the largest *producer* of oil. This meant a greater abundance of oil abroad than in the United States. Thus, after World War II, the major oil companies began maintaining two price levels: a domestic price for the United States and an international price. The domestic price was always higher, with the difference maintained by an import embargo on foreign oil. The embargo was repealed in the 1960s as oil reserves in the United States diminished.

There was such an excess of supply over demand in the international market that producers found it difficult to keep prices from dropping. Several of the oil-producing nations, of which most were located in the Middle East, formed a cartel known as the Organization of Petroleum Exporting Countries, or OPEC, in order to restrain competition and avoid excessive price drops. Meanwhile, the "Seven Sisters" petroleum companies also maintained strict limits on oil production in order to stabilize prices. But as the oil industry expanded, many independent oil companies were formed in the US and elsewhere — and those "independents" refused to limit production.

In 1959, a further element of instability entered the mix with the discovery in Libya of rich new reserves of high-quality, easily obtainable oil. Since it was located within easy access of the European market and the northeastern United States, both of which were major consuming areas, Libyan oil threatened the ability of the major oil companies to limit production and to prevent falling prices.

This implied threat became explicit when a new leader came to power in Libya in 1969. Colonel Muammar al-Qaddafi was unwilling to abide by the agreements between OPEC and the major oil companies. He soon nationalized most of the oil wells in Libya and took control of pricing. Occidental Petroleum, one of the largest independents, was primarily affected. Occidental sought help from Exxon, the biggest of the majors, believing that Qaddafi could be faced down by the withdrawal of oil experts from the country; however, in order to accomplish that, Occidental would need spare production to offset its Libyan losses. If it could obtain oil from Exxon at cost, it could afford to oppose the Libyan upstart. Exxon refused. Thus Qaddafi had asserted control over his nation's oil reserves while avoiding retaliation by the oil industry. As far as the majors were concerned, this set an unwelcome precedent. Moreover, Qaddafi drew the ire of the US government through his support for national liberation groups, such as the Palestine Liberation Organization (PLO), the Irish Republican Army (IRA), and the African National Congress (ANC).

By the 1960s, the United States was unable to produce as much oil as it was consuming, and began importing large quantities from other nations. This in itself provoked little concern. But in 1970 something truly extraordinary happened: America's rate of oil production peaked and began a long decline that has continued to the present. From this point onward, the US would become increasingly dependent on imported oil and would no longer be in a position unilaterally to stabilize world petroleum prices.

The major international oil companies periodically renegotiated the price of crude oil with the representatives of each exporting country. Most OPEC members were Middle-Eastern countries that opposed Israel in the ongoing Palestinian-Israeli conflict, while the US had been a staunch economic and military supporter of Israel since its creation in 1948. When Israel decisively defeated the principal Arab states in the 1967 war, the interim peace settlement left the Sinai peninsula in Israeli control. Egyptian President Sadat proposed a permanent peace if the Israelis returned all the occupied territories, but Israel refused. Since no progress was being made in negotiations, Sadat decided to initiate a war with limited objectives.

In October 1973, after demanding the evacuation of United Nations observers from the Egyptian-Israeli border, Egypt attacked Israeli forces in Sinai. Equipped with Soviet-provided surface-to-air missiles and armored vehicles, the Egyptian army overcame the ensuing Israeli air attack. However, the Israelis immediately requested and received replacement aircraft from the US. As the Israeli army prepared to cross the Suez Canal, a Soviet fleet moved into position in the eastern Mediterranean, raising the possibility that the war might escalate into a confrontation between the two superpowers. To avoid such a catastrophe, Israel and Egypt were pressured to accept a negotiated cease-fire. To the Israelis, this conflict became known as the Yom Kippur War, to the Arabs as the Ramadan War.

The Arab oil countries had been negotiating with the major oil companies when the war occurred. Because the war's outcome hinged on massive US military aid to Israel, the Arab states broke off negotiations and imposed an oil embargo against the US. An artificial oil shortage ensued. During the next four months, consumers in the United States were forced to wait in long lines at gas stations. The Arab members of OPEC effectively drove up crude oil prices fourfold. The OPEC cartel succeeded in wresting from the major oil companies the ability to set prices globally; from now on, it would be the producing countries, rather than the majors, that would play the key role in influencing the price of petroleum. However, while the embargo won certain benefits for OPEC members, higher oil prices also worked to the advantage of US oil companies.

A special relationship had existed between the US and Saudi Arabia since 1945, when FDR and Ibn Saud had concluded a pact ensuring secure oil exports in exchange for ongoing support for the Saudi regime. That relationship would become even more significant from 1973 on. As the world's largest oil exporter, Saudi Arabia would be in a position to dominate OPEC and to set prices. Both Washington and Riyadh would further cultivate their mutual interests, which over time would center on Saudi maintenance of global oil sales in US dollars, US arms sales to Saudi Arabia, Saudi investments of its oil revenues in the US, and Saudi control of world oil prices to American advantage.

Given the world's dependence on oil for transportation, industrial production, agriculture, and petroleum by-products, the 1973 price shock shattered the international economy and drove it into a period of inflation that would last until 1982. As the oil shortage reverberated through the global economy, the costs of industrial production and delivery of goods shot up. In 1974, the world experienced its greatest economic crisis since the 1930s. The period of substantial prosperity that had followed World War II came to an end.

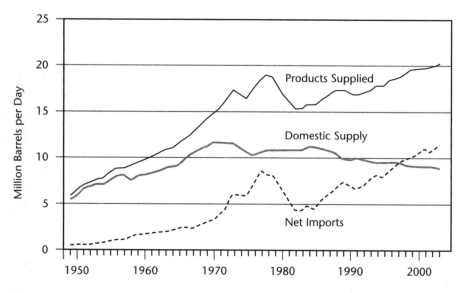

Figure 4. US petroleum overview, 1949-2003: US Trade in Petroleum and Petroleum Products
(Source: US Energy Information Administration)

In 1979, the Iranian people overthrew America's long-time brutal client, the Shah, who had been installed in office by the CIA in 1953. By this time, US and British leaders were bridling at the Shah's efforts to renegotiate his country's relationship with the oil companies doing business in Iran, and so Ayatolla Khomeini's rise to power was covertly supported. Soon war broke out between Iraq and Iran, two major oil-producing countries in the Middle East, with the US happy to see the countries mired in mutually destructive conflict. Since Europe imported large amounts of oil from both Iran and Iraq, a further artificial shortage ensued, and the international price of crude oil doubled again. In 1973, before the war between Egypt and Israel, oil prices had hovered at around $3 per barrel. After the embargo, the price rose to $12 per barrel, and after the commencement of the 1979 Iran-Iraq war it soared to more than $30 per barrel. As the resulting inflation worked its way through the international economy, the cost of all goods increased substantially. In the US, the price index of consumer goods rose approximately ten percent per year for several years in the late 1970s and early 1980s.

This economic upheaval motivated intense efforts toward energy conservation and the development of alternative energy sources (solar hot-water panels, wind turbines, methane digesters, etc). Industries made significant progress in improving building construction practices to conserve energy, in producing lighter and more fuel-efficient automobiles, and in developing more efficient lighting systems.

Investment in nuclear power plants increased as forecasters projected great increases in energy consumption in years to come. However, these projections went unrealized, due largely to the effectiveness of conservation measures. Conservation proved itself the least expensive response to the energy shortage. Nuclear power, by contrast, was extremely costly due to the expense of providing safety measures against the hazards of radiation release.

Higher oil prices also stimulated new exploration, which resulted in the discovery of new reserves in the North Sea, off the coasts of Nigeria and Angola, in Mexico, and on the north slope of Alaska.

Since the oil crises had not been due to a real shortage of oil, but to political events, conservation and increased reserves combined to generate a large oil surplus. OPEC was unable to sustain high prices, and in 1982 oil prices began to fall substantially. Inflation subsided and both the international economy and the US stock market enjoyed a recovery.

During the 1980s, many OPEC countries cheated on export quotas. Kuwait was the worst culprit, arbitrarily adding 50 percent to its reported reserves in

1985 to increase its quota, which was based on reserves. At the same time, Iran was motivated to cheat in order to finance its war with Iraq. Saudi Arabia found itself acting as the swing producer, reducing or increasing its production to keep prices stable; by 1985, the Saudis were selling less than half their quota. Up to this time, Britain had agreed to sell its North Sea oil at the official OPEC price, but Margaret Thatcher decided to let the price fluctuate. Reagan and Thatcher wanted to bring down the Soviets and persuaded King Fahd to increase his country's production substantially, thus dropping the price. The USSR relied on oil for foreign exchange, which it needed to match America's new arms buildup (principally, Star Wars).[10] In 1986, Saudi Arabia flooded the market and drove crude prices down sharply — from over $30 to less than $10; some producers were selling oil for as little as $6 a barrel. For the industry, this constituted a third oil shock: such low prices devastated US independent producers. Despite evident damage to the Soviet economy and despite Reagan's free-market rhetoric, George H.W. Bush — evidently acting at the behest of the US oil companies — persuaded the US administration to intervene and threaten to impose tariffs. Eventually both OPEC and non-OPEC producing countries agreed to coordinate production in order to maintain stable higher prices — though this never actually worked. The outcome was a victory for the Saudis, once again underscoring their power in the global market, but it led to a new commitment on America's part both to increase its presence in the Middle East (a decision that played a role in the lead-up to the Gulf War of 1990–91) and to diversify its import sources, relying more on Venezuela, Colombia, Ecuador, Canada, and Mexico, and less on the Arab states.

1980–2001: Lost Opportunities and the Prelude to Catastrophe

The oil crises of the 1970s had produced a significant shift in public attitudes about energy. Many groundbreaking books about the links between energy consumption and social and environmental problems were published during the late 1970s and early 1980s, including *Entropy*, by Jeremy Rifkin; *Soft Energy Paths*, by Amory Lovins; and *End of Affluence*, by Paul and Anne Ehrlich. President Jimmy Carter appeared on television to tell the American people that "[o]urs is the most wasteful nation on Earth; we waste more energy than we import" and to exhort the American people to engage in a massive national effort to conserve.[11] Spurred by tax subsidies and grants, businesses specializing in energy conservation, and in solar and wind power, sprang up by the hundreds.

With the advent of the Reagan-Bush administration in 1980, the official discourse on energy had suddenly changed again. In campaign commercials, Reagan's publicists proclaimed that it was "morning in America": the people of the US should forget their worries about energy-resource limits and return to their proper pastimes — spending, driving, and wasting. In a highly symbolic act, Reagan ordered the solar hot-water panels installed by Carter on the White House roof removed and junked. Subsidies for conservation measures and for the development and purchase of alternative-energy systems evaporated.

During the 1980s, financier Ivan Boesky proclaimed that "greed is good," and the US undertook massive investments in military hardware. The Reagan-Bush administration also covertly supported the Contras and other mercenary militias in Central America that opposed peasant efforts at land reform; significantly for future energy-related events, it also supported the Mujaheddin and other militant Islamist movements in Afghanistan, which opposed Soviet influence in south-central Asia. Representatives of the US administration convinced King Fahd of Saudi Arabia to pay for arms to be shipped from Egypt to the Mujaheddin, many of whom would later give rise to the Taliban and al Qaeda.

The collapse of the Soviet Union in 1991 — shortly following its oil-production peak in 1987 — came as a puzzling surprise to many US strategists, despite the fact that they had wished and plotted for this very eventuality for decades. Though the USSR had a long border with the Middle East, the US had managed to prevent the Soviets from forming strong trade or military alliances with any of the major Persian Gulf oil producers. The only major exception consisted of loans and trade agreements between the Soviets and Iraq. Had such alliances expanded, the USSR and the Middle East together would have had the resources necessary to successfully challenge the West, both economically and militarily. Moreover, the Soviets could have cushioned the effect of their oil-production peak in 1987 with imports, as the US did in the 1970s, and perhaps have avoided collapse. But this was not to be: as it happened, the USSR's production peak, coming shortly after the oil price drop of the mid-'80s, proved devastating to its oil export-dependent economy.

The US, with its former foe now in a state of economic chaos, found its ideological justifications for international military hegemony being undermined. Against whom or what was the US protecting the world now? Hence the

American search, beginning in the early 1990s, for new enemies to replace the old Soviet Union.

The Gulf War of 1991 began with a dispute between Iraq and Kuwait over ports in the Gulf and over oil export quotas. Iraq's plan to invade Kuwait appeared initially to have been condoned by the US, and so the real motives for the entrance of the United States into the conflict were unclear. President Bush (Sr.) had sent his emissary Henry Shuyler to persuade his then-ally Saddam Hussein to intervene in OPEC to hike oil prices for the benefit of his Texas constituents. Bush and his advisers knew that OPEC cheated and fell on the idea of a border incident whereby Iraq would take the southern end of the Rumaila field, from which the Kuwaitis were pumping. On the eve of the invasion, the US Ambassador in Baghdad, April Glaspie, said, "We have no opinion on the Arab-Arab conflicts like your border disagreement with Kuwait" — a statement countersigned by Secretary of State James Baker in Washington. Saddam assumed he had a wink and nod to invade his neighbor.[12]

Speculations about the reasons underlying the devastating US military intervention that followed had to do with Iraq's oil reserves, which were second only to those of Saudi Arabia — where the US installed permanent military bases during the war. In his book *Iraq and the International Oil System: Why America Went to War in the Gulf,* national security affairs analyst Stephen C. Pelletière examined US motives in the war in depth, concluding that the Gulf War represented a forcible expression of America's resolve to consolidate its control of the Middle East.[13] Iraq's decisive victory over Iran in 1988 had come as a shock to Washington, and neither the US nor Israel was about to tolerate a strong, independent, militarily competent Arab nation in the region.

Whatever the motive, the result was a quick military victory for the US and ongoing devastation for Iraq, which continued to suffer under UN-imposed trade sanctions throughout the following decade. During the hostilities, American strategists had apparently done some quick thinking and realized that they could make Saddam the swing producer of last recourse. By embargoing Iraq, they kept two to three million barrels of crude per day off the world market at no cost to anyone but Saddam. This is perhaps why they stopped at the gates of Baghdad and left him in power. Later, when oil prices rose uncomfortably, the US relaxed the embargo for "humanitarian" reasons, and most of Iraq's subsequent exports made their way to American gas tanks.

Beginning in 1993, the US attempted to control global oil prices through a policy of Dual Containment, in which the export quotas of Iraq and Iran were assigned to the nations of the lower Gulf as a reward for their economic and logistical support during the Gulf War. Thus one side effect of the war was that, because the US forcibly took most of Iraq's production off the world market for the ensuing decade, another world oil glut was averted — but only partly so. Despite a significant price surge during the war itself, the remainder of the decade saw stable though generally falling prices. Consequently, revenues to producing countries dwindled. Saudi Arabia, which had one of the fastest-growing populations in the world, faced diminishing per-capita incomes and simmering political unrest, the latter exacerbated by the presence of US military bases on land sacred to all Muslims.

Simultaneously with the development of the Dual Containment policy, the US began using the World Bank, the International Monetary Fund, and related institutions to secure non-OPEC oil sources through its funding of pipelines and exploration in less-consuming countries in Asia, Africa, and South America. From 1992 on, at least 21 agencies representing the American government, multilateral development banks, and other national governments approved billions of dollars in public financing in roughly 30 countries for energy projects that not only gave US oil companies new sources, but also aided a relatively new company called Enron — a complex entity involved in energy trading and distribution — to gain global reach. In India, Guatemala, Panama, Colombia, and other countries, Enron struck deals with local politicians giving the energy firm control over electrical and gas utilities. Enron executives began lavishing generous campaign donations on Democrats and Republicans alike, and soon US officials started twisting arms: for example, Mozambique was threatened with a cutoff of US foreign aid if it did not accept Enron's bid for a natural gas field.[14] Enron also implemented a domestic strategy, bribing state politicians to deregulate their utilities industries; in California, deregulation would result in the artificial energy crisis of 2001, in which utilities customers suffered through blackouts while the state itself racked up billions of dollars of debt in attempts to pay off electrical power generators and distributors (including Enron) that had been enabled by deregulation to systematically create both severe shortages and windfall profits.

The Clinton-Gore administration had taken office in 1993 amid high hopes on the part of environmentalists that some of the support for energy conservation and renewable energy programs that had flourished under Carter's administration would be revived. These hopes were based largely on Gore's timely book *Earth in the Balance* (1992). However, few substantial energy-policy changes were actually enacted during the following eight years.

Throughout the 1990s the most notable international political-economic development was the accelerating globalization of manufacturing, distribution, and corporate influence. Globalization was a complex phenomenon that outwardly had to do with high-speed communications, long-distance transportation, and the lowering of trade barriers through international agreements administered by trade adjudication bodies like the World Trade Organization; however, it had its roots in the inherent dynamics of industrialization itself.

In the previous two centuries, the machine-based production system had expanded by producing low-cost goods using fossil-fueled equipment to replace the skilled, and thus more expensive, labor of artisans. One result of industrialization was that the proportion of each enterprise's income going to wages typically fell, while the proportion going to investors and moneylenders gradually increased. This meant that, as more production processes became mechanized, the buying power of industrial workers would inevitably wane, resulting in a massive overproduction of goods and the bankrupting of the entire system, unless foreign markets could be found for manufactured products.

During the 20th century, more and more countries adopted mechanized production in the hope of escaping poverty. Those that had industrialized earlier were always in a favored position because they held both more economic power and also machine tools, dies, and patents to production processes; selling production rights and equipment to developing nations enriched the already industrialized countries. Slowly an industrial pyramid emerged. Though its apologists always dangled the promise that eventually the entire world would live at the same standard as people in Europe and America, in fact the pyramid was becoming steeper, with countries at the top growing richer and those at the bottom growing poorer. The same trend of increasing economic inequality was occurring within many countries as well, most notably the US.[15]

By the 1990s, the point had been reached where there were virtually no more pre-industrial markets to be taken over. This left corporations in the industrial pyramid with no one to displace but each other; inevitably, international competition grew much more intense.

Corporations adopted two strategies to survive: further automating, thus displacing human labor almost entirely; or moving production to countries where labor was cheaper. The combined result was that the share of industrial revenues being paid in wages and salaries fell even further, so that ever more people were left without the financial means to buy priced goods. For the hundreds of millions of people who had previously lived as self-sufficient rural peasants and who had been uprooted by the process of agricultural industrialization, the results were catastrophic. Corporations also began to merge at record rates since, lacking other new sources of revenue, they were now forced to consume each other.[16]

By this time the bulk of new investment capital was flowing not to manufacturing, but to speculation in fluctuating currencies, derivatives, options, and futures. The cumulative effect of such speculative investments was to enrich the financial elites and to undermine the long-term stability of the system as a whole. With surplus production capacity in almost every sector and investment capital leveraged to absurd lengths, the world teetered on the brink of economic collapse.

In the US, George W. Bush and Dick Cheney took office in 2001 following a deeply flawed election. With strong ties to the oil industry and to Enron, the new administration quickly proposed a national energy policy that focused on opening federally protected lands for oil exploration as well as on further subsidizing the oil industry; Cheney pointedly proclaimed that energy conservation "... may be a sign of personal virtue, but it is not a sufficient basis for a sound, comprehensive energy policy."[17]

Enron, George Bush's largest campaign contributor, had grown to become the 7th largest corporation in the US and the 16th largest in the world. Despite its reported massive profits, it had paid no taxes in four out of the five years from 1996 to 2001. The company had thousands of offshore partnerships, through which it had hidden over a billion dollars in debt. When this hidden debt was disclosed in October 2001, the company imploded. Its share price collapsed and its credit rating was slashed. Its executives resigned in disgrace, taking with them multimillion dollar bonuses, while employees and stockholders shouldered the immense financial loss. Enron's bankruptcy was the largest in corporate history up to that time, but its creative accounting practices appeared to be far from unique, with other corporations poised for a similar collapse.

In light of what was about to happen, the period from 1973 to 2001 can be seen as having represented a pivotal but lost opportunity. The oil embargo of 1973 and the global economic turmoil accompanying the Iranian revolution made it clear how dependent the world economy had become upon petroleum, and how dependent the US had become upon oil imports. Moreover, everyone knew that oil was a resource that was inherently nonrenewable and therefore limited in supply. The rational response would have been to undertake massive, ongoing conservation efforts and investments in a transition to renewable energy sources. Such efforts were tentatively begun, but quickly abandoned. Greed and political influence on the part of the oil companies were no doubt factors in preventing that course from being pursued. But the companies do not deserve all of the blame: free-market economists and their acolytes in political office genuinely believed that the all-knowing market would provide for every contingency and that resource shortages would never amount to a serious problem.

In hindsight, the reasons for abandoning the path of conservation seem tragically wrongheaded. There was at the time a sizeable minority who decried the return to heedless consumerism, but their voice was destined not to prevail. The path actually taken was one not only of consumptive excess and increased global competition but also of a growing attempt on the part of US geopolitical strategists to control global petroleum resources. Its consequences would materialize dramatically on the morning of September 11, 2001.

3

Lights Out: Approaching the Historic Interval's End

Pangloss is admired, and Cassandra is despised and ignored. But as the Trojans were to learn to their sorrow, Cassandra was right, and had she been heeded, the toil of appropriate preparation for the coming adversity would have been insignificant measured against the devastation that followed a brief season of blissful and ignorant optimism ....

Today, Cassandra holds advanced degrees in biology, ecology, climatology, and other theoretical and applied environmental sciences. In a vast library of published book and papers, these scientists warn us that if civilization continues on its present course, unspeakable devastation awaits us or our near descendants ....

As a discomforted public, and their chosen political leaders, cry out "Say it isn't so!", there is no shortage of reassuring optimists to tell us, "Don't worry, be happy."

We sincerely wish that we could believe them. But brute scientific facts, and the weakness of the Panglossian arguments, forbid.

— Ernest Partridge (2000)

... by early in the twenty-first century, the era of pumping "black gold" out of the ground to fuel industrial societies will be coming to an end.

— Paul Ehrlich (1974)

We've embarked on the beginning of the last days of the age of oil.

— Mike Bowlin, Chairman and CEO, ARCO (1999)

My father rode a camel. I drive a car. My son flies a jet airplane. His son will ride a camel.

— Saudi saying

85

The September 11 atrocities so dominated world news, politics, military affairs, and the economy that popular discussion soon divided all of recent history into two categories: "pre-9/11" and "post-9/11." For most Americans, the events were not only horrifying, but also entirely unexpected. Given the reputed Middle-Eastern origin of the airplane hijackers, many people suspected that oil was somehow involved.

Of the hijackers themselves, fifteen out of nineteen were described as Saudi Arabian nationals. American officials identified the mastermind of the attacks as Osama bin Laden, a scion of one of Saudi Arabia's wealthiest families — a family that had long-standing financial ties to the Bushes (the bin Ladens had, via an intermediary, helped finance George W. Bush's first business venture, Arbusto Energy Company, in 1979). Osama, according to his own published statements, regarded US military bases in Saudi Arabia as an affront to Islam.

As the world's largest oil producer, the Saudi kingdom had remained a faithful US client for decades, but a growing, youthful population and diminishing oil revenues were beginning to generate popular unrest within that nation. The Saudi royal family had sought to defuse any possible Islamist opposition by officially backing the ultraconservative Wahaabi sect and by permitting considerable sums in petrodollars to go to the financing of radical — some would say terrorist — Islamist groups both within and beyond the country's borders. To these latter efforts, recent American administrations had agreed to turn a blind eye in return for continued Saudi cooperation in maintaining stable oil prices.

For its part, the American leadership had been manipulating radical Islamist movements for decades. In Afghanistan during the 1980s, and the Balkans and Chechnya in the '90s, the US secretly armed and funded Islamist terror networks in order to destabilize troublesome nations. That tactic had achieved spectacular success against the USSR, whose disastrous military efforts to maintain control of neighboring Afghanistan constituted one of the principal factors leading to the Soviet empire's downfall. However, the radical Islamists, though willing to accept guns and dollars, had no natural sympathy for American interests.

Somewhat ingeniously, starting in the mid-1990s, the US intelligence community used the Islamists' obstreperousness to its advantage by dangling the threat of radical Muslim "terrorism"[1] before its domestic audience as a way of gaining support for increased military and security budgets and of obtaining ever greater authority for surveillance, extrajudicial detentions, and other suspensions of civil liberties.

Osama bin Laden had been a key figure in the American-supported militant Islamist movement throughout most of the 1980s. Exactly when the US ceased indirectly sponsoring his activities is unclear. Libya was the first nation to call for his arrest, in 1994. Following two terrorist attacks against US interests later in the 1990s, after which then-President Clinton posted a reward for bin Laden's capture, the latter moved his headquarters to Afghanistan, where he trained his al Qaeda agents in secret bases — many of which had been planned or built by the CIA during the 1980s.

The Bush administration's response to the 9/11 attacks was to bomb Afghanistan, remove the Taliban regime from power, and install a compliant interim client government in its place.

A few commentators pointed out that Afghanistan was located near the strategically significant oil and gas reserves of the Caspian Sea, speculating that the war might be an effort to enforce the building of a gas pipeline through Afghanistan to warm-water ports in Pakistan. Two French investigative journalists, Jean-Charles Brisard and Guillaume Dasquie, even claimed that the US action in Afghanistan had been contemplated — if not planned — for months prior to the 9/11 attacks, as indicated by threats purportedly made to Taliban representatives during the pipeline negotiations. According to Brisard and Dasquie, in a meeting in Islamabad in August 2001 between Christina Rocca, in charge of Central Asian affairs for the US Government, and the Taliban ambassador to Pakistan, the Taliban were told, "either you accept our offer of a carpet of gold, or we bury you under a carpet of bombs."[2]

Others, including some oil-industry insiders, disputed the idea that the war was essentially about oil or natural gas, pointing out that Afghanistan was not itself essential to the domination of energy resources in the region and that the proposed pipeline was of minor economic consequence to the US. Thus both the ostensible and the real US motives must have been simply the pursuit of bin Laden and his organization.

While most people seemed to find these latter arguments convincing, a few important points should be emphasized: If not for oil, the US would have little interest in the Middle East. If not for US involvement in the Middle East (specifically, Saudi Arabia), Osama bin Laden would never have felt compelled to destroy symbols of American economic and military power. In this respect, though the violence took place in Afghanistan, New York, and Washington, the real strategic objectives on both sides had much to do with Saudi Arabia. Moreover, it appeared that pre-9/11 investigations by the FBI into Al Qaeda had been systematically obstructed by orders from the highest levels of the US government, perhaps to divert attention away from certain members of the Saudi royal family and the bin Laden family, who had for years been financially supporting Osama bin Laden.[3]

Thus energy resources lay at the heart of the conflict in any case. Further, however, the Afghan war entailed the construction of permanent American military bases throughout Central Asia — which, if US leaders had indeed determined to control the future exploitation of the oil and gas resources of the Caspian, would be of considerable assistance in that effort.

The Bush administration quickly proclaimed that the Afghanistan campaign was only the beginning of its "war on terrorism," and officials floated lists of other potential targets, numbering from three to nearly fifty nations. Critics of the Bush policy claimed that the administration had, in effect, declared war on much of the rest of the world. Most of the listed nations possessed important oil resources while many — including Iran and Iraq, which were high on the lists — had little or no discernible relationship with bin Laden or Al Qaeda. With "terrorism" as its ostensible but elusive enemy, the American administration appeared to be embarking on a grandiose plan to use its military might to gain footholds in strategic regions around the globe, and perhaps to seize full and direct control of the world's petroleum resources.

On to Mesopotamia

Soon after the invasion of Afghanistan, the Bush administration turned its attention toward Iraq, claiming that the country was in defiance of UN resolutions and that it was harboring or collaborating with the perpetrators of the 9/11 atrocities. In response Saddam Hussein re-admitted UN inspectors, who began scouring the countryside for banned weapons. Bush administration officials continued escalating their rhetoric. In the words of Vice President Dick Cheney, "there is no doubt that Saddam Hussein now has weapons of mass destruction." A preemptive invasion would be justified and necessary because, according to Condoleeza Rice, "we don't want the smoking gun to be a mushroom cloud." Cheney also sought to tie the Iraqi regime with the 9/11 perpetrators, calling Baghdad a safe haven for terrorists. US officials made it clear that, with or without international backing, they intended to invade Iraq and depose its leader.

Millions of citizens in nations around the world took to the streets in unprecedented numbers to voice their dismay at the prospect of this preemptive attack.

Washington officials tried but failed to obtain a UN Security Council resolution specifically authorizing the invasion, since without such a resolution the contemplated military action would be illegal under the UN charter (as was noted months later by UN Secretary General Kofi Annan). Opposition prinicipally from France, Germany and Russia prevented passage of the resolution. In the course of the debate, long-standing alliances frayed.

The US, with the participation of Great Britain and token help from several other countries, launched the invasion in March 2003. The immediate military operations proceeded relatively quickly (concluding in a mere six weeks), though at points supply lines became so dangerously stretched that it was clear Washington had committed too few troops to do the job properly. Iraqis at first defended their homeland and thousands died. However, as US troops encircled Baghdad, resistance suddenly and mysteriously vanished. Soon the world was viewing video footage of a staged demonstration in which a small, handpicked group of Iraqis — aided by US soldiers in tanks and an armored personnel carrier — pulled down, defaced, and danced on a statue of Saddam Hussein.

Looters then began removing the archaeological and artistic treasures of the oldest civilization on Earth from Iraq's national museums and galleries. Though the Geneva Conventions clearly state that it is the responsibility of occupying forces to protect the lives and property of occupied peoples, US officials shrugged. War Secretary Rumsfeld mused that the looters were merely "blowing off steam." "It's untidy," Rumsfeld opined, "and freedom's untidy. And free people are free to make mistakes and commit crimes." By mid-April there was nothing left to loot — except at the Oil Ministry, the only government building to have been guarded by US troops.

American forces had also aggressively taken control of — and defended — the Iraqi oil fields, ports, and refineries.

Many Iraqis did initially rejoice at the overthrow of their hated dictator. However, resistance to the occupation materialized and expanded over the following months. As US casualties mounted week by agonizing week, it became clear that the entire exercise had been steeped in catastrophic miscalculation. Despite warnings from the CIA and the State Department, the war planners had assumed an easy post-invasion transition to capitalist democracy. But now, as Iraqi unemployment soared, and as Sunnis and Shias carried out dozens of daily attacks against the occupying forces, the nation teetered on the brink of civil war.

By October 2004, US combat deaths numbered over a 1000, with the Iraqi civilian toll between 10 and a 100 times that number. Unexploded cluster bombs littered the nation, and depleted uranium — from US and British tank shells, bombs, and bullets — was causing soaring rates of cancers and birth defects. Basic services were still only partially restored, and large sections of the country were too dangerous for any but the most intrepid journalist to visit. Polls showed that the vast majority of Iraqis wanted the US to leave immediately.

Also by this time, the US government had officially admitted that Saddam Hussein had been telling the truth when he said he had no banned weapons, these having been destroyed soon after the first Gulf War. The Bush administration blamed faulty intelligence, saying that its officials genuinely believed that the weapons existed, and implied that this misreading of the situation was the CIA's fault. However, CIA officials chafed and pointed to clear documentation showing that the agency had told the administration before the invasion that claims being made publicly about Iraq's imminent development of nuclear weapons, and about links between Saddam Hussein and al Qaeda, were exaggerated.

Still other documents implied that the invasion had been planned months in advance, even prior to 9/11.

The Iraq and Afghanistan invasions were also accompanied by the extraordinary detention of individuals suspected of having some link with "terrorism." Suspects were swept up and held without charge or trial, and denied the right to seek legal counsel, or even to telephone relatives. Media and human rights organizations received no lists of detainees. Later it would emerge that many detainees in Iraq, Afghanistan, and the Guantanamo Bay detention facility in Cuba were being tortured.

During the first months after 9/11, the American media devoted little airtime to those questioning official policy. Television programs revolved around patriotic themes; flags sprouted on TV screens and billboards. But gradually, tentatively at first, public expressions of dissent began to emerge in print and in a remarkably popular string of video documentaries from various independent producers. By the election season of 2004, it had become apparent that the nation was deeply polarized. While many believed that everything being done by the government in its "war on terror" was justified, others were convinced that the administration's actions were both criminal and incompetent.

As all of this played out, the new Department of Homeland Security issued repeated warnings of further terrorist attacks. The US military had failed so far to locate, capture, or kill Osama bin Laden, while the US presence in Iraq appeared to be providing a powerful recruiting tool for Islamist militant organizations.

Yet al Qaeda, the organization that was purportedly organizing terrorist actions around the world, remained shadowy. Although the US claimed to have captured or killed most of the top leadership, not a single confirmed al Qaeda member had been convicted of a crime. A BBC documentary series ("The Power of Nightmares: The Rise of the Politics of Fear") aired in late 2004 went so far as to claim that al Qaeda was not an organized international network at all, that it did not have members or a leader, nor did it have "sleeper cells" or an overall strategy. In fact, according to series producer Adam Curtis, al Qaeda barely even existed, except as an idea about cleansing a corrupt world through religious violence.

But if al Qaeda was a mirage, then who was responsible for the events of September 11, 2001? As those events passed into history, questions about what had actually happened that day only deepened. The Bush administration had destroyed physical evidence and delayed and inhibited an inquiry. When an official inquiry was eventually convened, the administration was able to handpick its key members. Important questions went unasked. Many relatives of the 9/11 victims, among tens of thousands of others, began to conclude that the available evidence suggested some form of government complicity in the events — at a minimum, willful and knowing efforts to hamper investigative efforts that could have prevented the attacks.

In short, during the first term of the Bush administration, something extraordinary had happened. The US had, it appeared, manufactured new enemies to replace the Soviet Union. And it was using these new enemies to justify the invasion and occupation of nations in the Middle East and Central Asia and the building of large, permanent military bases in those regions. Old alliances were being broken; new lines of contention were emerging.

If the Iraq invasion had never had anything to do with weapons of mass destruction or al Qaeda, it did seem to have something to do with oil. But what, exactly? The US had not simply commandeered Iraq's oil supplies. In fact, oil was flowing less freely from that country than it had prior to the invasion. What could Washington have hoped to accomplish?

The beginning of an answer to that question emerged in 2004, as oil prices began a long and steep climb. On previous occasions when oil prices spiked (in 1973, 1979, and 1991) there had been easy explanations. Now analysts cited many converging factors — strikes in Norway, political unrest in Nigeria, pipeline sabotage in Iraq — that, even when added together, hardly seemed to justify the dramatic price surge.

The actual cost of extracting oil had not increased substantially. If the high prices continued, importing nations like the US would be squeezed, while exporters would luxuriate in new profits.

In the 1970s, as well, immense wealth had flowed from oil-importing nations to oil exporters. The US had been able to control the situation then by getting the oil exporters to reinvest their newfound wealth in the US economy. This was not a hard sell, as at that time the US offered investors a universally accepted currency and the world's most stable economy.

But now the situation was different. After decades of increasing trade deficits and three years of exploding national debt, the US economy was increasingly an empty shell. And the US dollar now had a rival for the title of world reserve currency — the euro. It seemed that America might soon lose its ability to control the flow of wealth around the world. Yet it still held one ace in hand: its extraordinary military machine.

The US military seemed to be acting as a global oil cop. Almost everywhere there were significant oil and gas pipelines or fields, one could expect to find a US base close by. American troops were stationed in 120 countries, now including most of the major oil producers (with the remaining notable exceptions of Iran and Russia). The US appeared to be the center of an empire of oil, facing an elusive enemy whose unpredictable attacks were capable of justifying intervention anywhere, at any time.

Was the price runup of 2004 in any way predictable? Was there some deeper reason for it? And could American officials' ability to foresee major oil price increases, with disastrous eventual impacts, have led them to undertake desperate measures?

The Ground Giving Way

In nearly every year since 1859, the total amount of oil extracted from the world's ancient and finite underground reserves had grown — from a few thousand barrels a year to 65 million barrels per day by the end of the 20th century, an increase averaging about two percent per annum. Demand had grown just as dramatically, sometimes lagging behind the erratically expanding supply. The great oil crises of the 1970s — the most significant occasions when demand exceeded supply — had been politically-based interruptions in the delivery of crude that was otherwise readily available; there had been no actual physical shortage of the substance then, or at any other time.

In the latter part of the year 2000, as Al Gore and George W. Bush were criss-crossing the nation vying for votes and campaign contributions, the world price of oil rose dramatically from its low point of $10 per barrel in February 1999 to $35 per barrel by mid-September of 2000. Essentially, Venezuela and Mexico had convinced the other members of OPEC to cease cheating on production quotas, and this resulted in a partial closing of the global petroleum spigot. Yet while Saudi Arabia, Iraq, and Russia still had excess production capacity that could have been brought on line to keep prices down, most other oil-producing nations were pumping at, or nearly at, full capacity throughout this period.

Meanwhile, a wave of mergers had swept the industry. Exxon and Mobil had combined into Exxon-Mobil, the world's largest oil company; Chevron had merged with Texaco; Conoco had merged with Phillips; and BP had purchased Amoco-Arco. Small and medium-sized companies — such as Tosco, Valero, and Ultramar Diamond Shamrock Corp — also joined in the mania for mergers, buyouts, and downsizing. Nationally, oil-company mergers, acquisitions, and divestments totaled $82 billion in 1998 and over $50 billion in 1999.

Altogether, the oil industry appeared to be in a mode of consolidation, not one of expansion. As Goldman Sachs put it in an August 1999 report, "The oil companies are not going to keep rigs employed to drill dry holes. They know it but are unable ... to admit it. The great merger mania is nothing more than a scaling down of a dying industry in recognition that 90 percent of global conventional oil has already been found."[4]

Meanwhile the Energy Information Agency (EIA) predicted that global *demand* for oil would continue to grow, increasing 60 percent by the year 2020 to roughly 40 billion barrels per year, or nearly 120 million barrels per day.[5]

The dramatic price hikes of 2000 soon triggered a global economic recession. The link between energy prices and the economy was intuitively obvious and had been amply demonstrated by the oil crises and accompanying recessions of the 1970s. Yet, as late as mid-2000, many pundits were insisting that the new "information economy" of the 1990s was impervious to energy-price shocks. This trend of thought was typified in a comment by British Prime Minister Tony Blair, who in January 2000 stated that "[t]wenty years on from the oil shock of the '70s, most economists would agree that oil is no longer the most important commodity in the world economy. Now, that commodity is infor-mation."[6] Yet when fuel prices soared in Britain during the last quarter of the year, truckers went on strike, bringing commerce within that nation to a virtual standstill. Though energy resources now *directly* accounted for only a small portion of economic activity in industrialized countries — 1.2 percent to 2 percent in the US — all manufacturing and transportation still required fuel. In fact, the *entire* economy in every industrial nation was completely dependent on the continuing availability of energy resources at low and stable prices.

As the world economy slowed, demand for new goods also slowed, and manufacturing and transportation were scaled back. As a result, demand for oil also decreased, falling roughly five percent in the ensuing year. Prices for crude began to soften. Indeed, by late 2001, oil prices had plummeted partly as the result of market-share competition between Russia and Saudi Arabia. Gasoline prices at the pump in California had topped $2 in late 2000, but by early 2002 they had drifted to a mere $1.12 per gallon.

Such low prices tended to breed complacency. The Bush administration warned of future energy shortages, but proposed to solve the problem by promoting exploration and production within the US and by building more nuclear power plants — ideas that few with much knowledge of the energy industry took seriously. Now that gasoline prices were low again, not many citizens contemplated the possible future implications of the price run-ups of 2000. In contrast, industry insiders expressed growing concern that fundamental limits to oil production were within sight.

This concern gained public recognition in 2004, as oil prices again shot upward, this time attaining all-time highs of over $55 per barrel. *National Geographic* proclaimed in its cover story that this was "The End of Cheap Oil"; *Le Monde* announced "The Petro-Apocalypse;" while Paul Erdman, writing for the CBS television magazine *Marketwatch,* proclaimed that "the looming oil crisis will dwarf 1973." In article after article, analysts pointed to dwindling discoveries of new oil, evaporating spare production capacity, and burgeoning global demand for crude. The upshot: world oil production might be near its all-time peak.

If this were indeed the case — that world petroleum production would soon no longer be able to keep up with demand — it would be the most important news item of the dawning century, dwarfing even the atrocities of September 11. Oil was what had made 20th-century industrialism possible; it was the crucial material that had given the US its economic and technological edge during the first two-thirds of the century, enabling it to become the world's superpower. If world production of oil could no longer expand, the global economy would be structurally imperiled. The implications were staggering.

There was every reason to assume that the Bush administration understood at least the essential outlines of the situation. Not only were many policy makers themselves — including the President, Vice President, and National Security Advisor — former oil industry executives; in addition, Vice President Dick Cheney's chief petroleum-futures guru, Matthew Simmons, had warned his clients of coming energy-supply crises repeatedly. Moreover, for many years the CIA had been monitoring global petroleum supplies; it had, for example, subscribed to the yearly report of Switzerland-based Petroconsultants, published at $35,000 per copy, and was thus surely also aware of another report, also supplied by Petroconsultants, titled "The World's Oil Supply 1995," which predicted that the peak of global oil production would occur during the first decade of the new century.

It would be an understatement to say that the general public was poorly prepared to understand this information or to appreciate its gravity. The *New York Times* had carried the stories of the oil company mergers on its front pages, but offered its readers little analysis of the state of the industry or that of the geological resources on which it depended. Mass-audience magazines *Discover* and *Popular Science* blandly noted, in buried paragraphs or sidebars, that "early in [the new century] ... half the world's known oil supply will have been used, and oil production will slide into permanent decline"[7] and that "experts predict that production will peak in 2010, and then drop over subsequent years"[8] — but these publications made no attempt to inform readers of the monumental implications of these statements. It would be safe to say that the average person had no clue whatever that the entire world was poised on the brink of an economic cataclysm that was as vast and unprecedented as it was inevitable.

Yet here and there were individuals who did perfectly comprehend the situation. Many were petroleum geologists who had spent their careers searching the globe for oil deposits, honing the theoretical and technical skills that enabled them to assess fairly accurately just how much oil was left in the ground, where it was located, and how easily it could be accessed.

What these people knew about the coming production peak — and how and when they arrived at this knowledge — constitutes a story that centers on the work of one extraordinary scientist.

M. King Hubbert: Energy Visionary

During the 1950s, '60s, and '70s, Marion King Hubbert became one of the best-known geophysicists in the world because of his disturbing prediction, first announced in 1949, that the fossil-fuel era would prove to be very brief.

Of course, the idea that oil would run out eventually was not, in itself, original. Indeed, in the 1920s many geologists had warned that world petroleum supplies would be exhausted in a matter of years. After all, the early wells in Pennsylvania had played out quickly; and extrapolating that initial experience to the limited reserves known in the first two decades of the century yielded an extremely pessimistic forecast for oil's future. However, the huge discoveries of the 1930s in east Texas and the Persian Gulf made such predictions laughable. Each year far more oil was being found than was being extracted. The doomsayers having been proven wrong, most people associated with the industry came to assume that supply and demand could continue to increase far into the future, with no end in sight. Hubbert, armed with better data and methods, doggedly challenged that assumption.

M. King Hubbert had been born in 1903 in central Texas, the hub of world oil exploration during the early 20th century. After showing a childhood fascination with steam engines and telephones, he settled on a career in science. He earned BS, MS, and Ph.D. degrees at the University of Chicago and, during the 1930s, taught geophysics at Columbia University. In the summer months, he worked for the Amerada Petroleum Corporation in Oklahoma, the Illinois State Geological Survey, and the United States Geological Survey (USGS). In 1943, after serving as a senior analyst at the Board of Economic Warfare in Washington, DC, Hubbert joined Shell Oil Company in Houston, where he directed the Shell research laboratory. He retired from Shell in 1964, then joined the USGS as a senior research geophysicist, a position he held until 1976. In his later years, he also taught occasionally at Stanford University, the University of California at Los Angeles, the University of California at Berkeley, the Massachusetts Institute of Technology, and Johns Hopkins University.

During his career, Hubbert made many important contributions to geophysics. In 1937 he resolved a standing paradox regarding the apparent strength of rocks that form the Earth's crust. Despite their evident properties of hardness and brittleness, such rocks often show signs of plastic flow. Hubbert demonstrated mathematically that, because even the hardest of rocks are subject to immense pressures at depth, they can respond in a manner similar to soft muds or clays. In the early 1950s, he showed that underground fluids can become entrapped under circumstances previously not thought possible, a finding that resulted in the redesign of techniques employed to locate oil and natural gas deposits. And by 1959, in collaboration with USGS geologist William W. Rubey, Hubbert also explained some puzzling characteristics of overthrust faults — low-angle fractures in rock formations in which one surface is displaced relative to another by a distance on the order of kilometers.

These scientific achievements would have been sufficient to assure Hubbert a prominent place in the history of geology. However, his greatest recognition came from his studies of petroleum and natural gas reserves — studies he had begun in 1926 while a student at the University of Chicago. In 1949, he used statistical and physical methods to calculate total world oil and natural gas supplies and documented their sharply increasing consumption. Then, in 1956, on the basis of his reserve estimates and his study of the lifetime production profile of typical oil reservoirs, he predicted that the peak of crude-oil production in the United States would occur between 1966 and 1972. At the time, most economists, oil companies, and government agencies (including the USGS) dismissed the prediction. The actual peak of US oil production occurred in 1970, though this was not apparent until 1971.[9]

Let us trace just how Hubbert arrived at his prediction. First, he noted that production from a typical reservoir or province does not begin, increase to some stable level, continue at that level for a long period, and then suddenly drop off to nothing after all of the oil is gone. Rather, production tends to follow a bell-shaped curve. The first exploratory well that punctures a reservoir is capable of extracting only a limited amount; but once the reservoir has been mapped, more wells can be drilled.

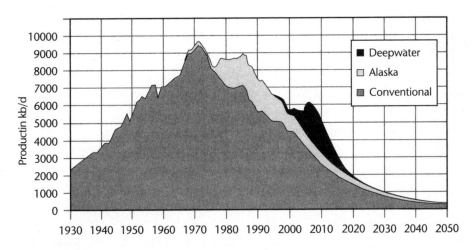

Figure 5. US oil production, history and projection, including lower 48, Alaska and Gulf of Mexico (deep water). Source: ASPO.

During this early phase, production increases rapidly as the easiest-accessed oil is drained first. However, beyond a certain point, whatever remains is harder to get at. Production begins to decline, even if more wells are still being drilled. Typically, the production peak will occur when about half of the total oil in the reservoir has been extracted. Even after production has tapered off, some oil will still be left in the ground: it is economically impractical — and physically impossible — to remove every last drop. Indeed, for some reservoirs only a few percent of the existing oil may be recoverable (the average is between 30 and 50 percent).

Hubbert also examined the history of discovery in the lower-48 United States. More oil had been found in the 1930s than in any decade before or since — and this despite the fact that investment in exploration had increased dramatically in succeeding decades. Thus discovery also appeared to follow a bell-shaped curve. Once the history of discovery had been charted, Hubbert was able to estimate the total ultimately recoverable reserves (URR) for the entire lower-48 region. He arrived at two figures: the most pessimistic reasonable amount (150 billion barrels) and the most optimistic reasonable amount (200 billion barrels). Using these two estimates, he calculated future production rates. If the total URR in the lower-48 US amounted to 150 billion barrels, half would be gone — and production would peak — in 1966; if the figure were closer to 200 billion barrels, the peak would come in 1972.

These early calculations involved a certain amount of guesswork. For example, Hubbert chose to chart production rates on a logistic curve, whereas he might have employed a better-fitting Gaussian curve.[10] Even today, according to Princeton University geophysicist Kenneth S. Deffeyes, author of *Hubbert's Peak: The Impending World Oil Shortage*, the "numerical methods that Hubbert used to make his prediction are not crystal clear."[11] Despite many conversations with Hubbert and ensuing years spent attempting to reconstruct those original calculations, Deffeyes finds aspects of Hubbert's process obscure and "messy." Nevertheless, Hubbert did succeed in obtaining important, useful findings.

Following his prediction of the US production peak, Hubbert devoted his efforts to forecasting the global production peak. With the figures then available for the likely total recoverable world petroleum reserves, he estimated that the peak would come between the years 1990 and 2000. This forecast would prove too pessimistic, partly because of inadequate data and partly because of minor flaws in Hubbert's method. Nevertheless, as we will see shortly, other researchers would later refine both input data and method in order to arrive at more reliable predictions — ones that would vary only about a decade from Hubbert's.

Hubbert immediately grasped the vast economic and social implications of this information. He understood the role of fossil fuels in the creation of the modern industrial world, and thus foresaw the wrenching transition that would likely occur following the peak in global extraction rates. In lectures and articles, starting in the 1950s, Hubbert outlined how society needed to change in order to prepare for a post-petroleum regime. The following passage, part of a summary by Hubbert of one of his own lectures, conveys some of the breadth and flavor of his macrosocial thinking:

> The world's present industrial civilization is handicapped by the coexistence of two universal, overlapping, and incompatible intellectual systems: the accumulated knowledge of the last four centuries of the properties and interrelationships of matter and energy; and the associated monetary culture which has evolved from folkways of prehistoric origin.
>
> The first of these two systems has been responsible for the spectacular rise, principally during the last two centuries, of the present industrial system and is essential for its continuance. The second, an

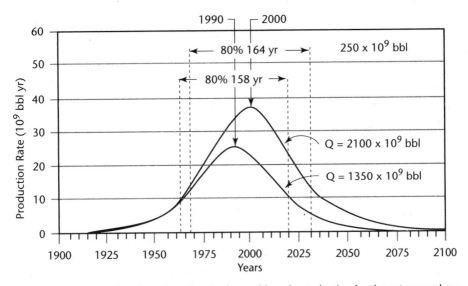

Figure 6. M. King Hubbert's projected cycles for world crude production for the extreme values of the estimated total resource. (Source: M. K. Hubbert, Resources and Man)

inheritance from the prescientific past, operates by rules of its own having little in common with those of the matter-energy system. Nevertheless, the monetary system, by means of a loose coupling, exercises a general control over the matter-energy system upon which it is superimposed.

Despite their inherent incompatibilities, these two systems during the last two centuries have had one fundamental characteristic in common, namely exponential growth, which has made a reasonably stable coexistence possible. But, for various reasons, it is impossible for the matter-energy system to sustain exponential growth for more than a few tens of doublings, and this phase is by now almost over. The monetary system has no such constraints, and, according to one of its most fundamental rules, it must continue to grow by compound interest.[12]

Hubbert thus believed that society, if it is to avoid chaos during the energy decline, must give up its antiquated, debt-and-interest-based monetary system and adopt a system of accounts based on matter-energy — an inherently ecological system that would acknowledge the finite nature of essential resources.

Hubbert was quoted as saying that we are in a "crisis in the evolution of human society. It's unique to both human and geologic history. It has never happened before and it can't possibly happen again. You can only use oil once. You can only use metals once. Soon all the oil is going to be burned and all the metals mined and scattered."[13]

Statements like this one gave Hubbert the popular image of a doomsayer. Yet he was not a pessimist; indeed, on occasion he could assume the role of utopian seer. We have, he believed, the necessary know-how; all we need do is overhaul our culture and find an alternative to money. If society were to develop solar-energy technologies, reduce its population and its demands on resources, and develop a steady-state economy to replace the present one based on unending growth, our species' future could be rosy indeed. "We are not starting from zero," he emphasized. "We have an enormous amount of existing technical knowledge. It's just a matter of putting it all together. We still have great flexibility but our maneuverability will diminish with time."[14]

Reading Hubbert's few published works — for example, his statement before the House of Representatives Subcommittee on the Environment on June 6, 1974 — one is struck by his ability to follow the implications of his findings on oil depletion through the domains of economics and ecology.[15] He was a holistic and interdisciplinary thinker who deserves, if anyone does, to be called a prophet of the coming era.

Hubbert died in 1989, a few years before his predicted date for the global production peak. That all-important forecast date was incorrect, as the rate of world oil production continued to increase through the first months of 2005. But by how far did he miss the mark? It would be up to his followers to find out.

Hubbert's Legacy

Since Hubbert's death, several other prominent petroleum geologists have used their own versions of his method to make updated predictions of the world's oil production peak. Their results diverge only narrowly from one another's. Since these scientists have been able to maintain updated data on reserves and production rates and since their work figures prominently in the current discussion about petroleum depletion, it will be helpful to introduce some of these individuals.

Colin J. Campbell is by most accounts the dean among Hubbert's followers. After earning his Ph.D. at Oxford in 1957, Campbell worked first for Texaco and then Amoco as an exploration geologist, his career taking him to Borneo, Trinidad, Colombia, Australia, Papua New Guinea, the US, Ecuador, the United Kingdom, Ireland, and Norway. He later was associated with Petroconsultants in Geneva, Switzerland, and in 2001 brought about the creation of the Association for the Study of Peak Oil (ASPO), which has members affiliated with universities in Europe. He has published extensively on the subject of petroleum depletion, and is author of the book *The Coming Oil Crisis*.[16]

Campbell's most prominent and influential publication was the article "The End of Cheap Oil?", which appeared in the March 1998 issue of *Scientific American*. The co-author of that article, Jean Laherrère, had worked for the oil company Total (now Total Fina Elf) for thirty-seven years in a variety of roles encompassing exploration activities in the Sahara, Australia, Canada, and Paris. Like Campbell, Laherrère had also been associated with Petroconsultants in Geneva.

The *Scientific American* article's most arresting features were its sobering title and its conclusion:

> From an economic perspective, when the world runs completely out of oil is ... not directly relevant: what matters is when production

begins to taper off. Beyond that point, prices will rise unless demand declines commensurately. Using several different techniques to estimate the current reserves of conventional oil and the amount still left to be discovered, we conclude that the decline will begin before 2010.[17]

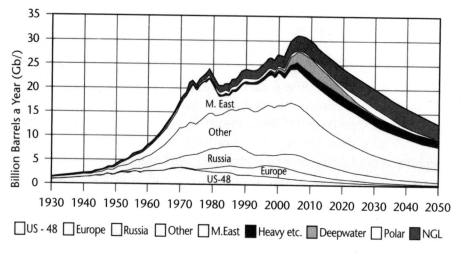

Figure 7a. World oil production, history and projection. (Source: ASPO)

Amount		Gb	Annual Rate - Regular Oil					Gb	Peak
Regular Oil			Mb/d	2005	2010	2020	2050	Total	Date
Past	**Future**	**Total**	US-48	3.4	2.7	1.7	0.4	200	1972
Known Fields	New		Europe	5.2	3.6	1.8	0.3	75	2000
945	770	135	1850	Russia	9.1	8	5.4	1.5	210	1987
905			ME Gulf	20	20	20	12	675	1974
All Liquids			Other	29	25	17	8	690	2004
1040	1360	2400	**World**	**66**	**60**	**46**	**22**	**1850**	2006
2004 Base Scenario			**Annual Rate - Other**						
M. East producing at capacity (anomalous reporting corrected)			Heavy etc.	2.4	4	5	4	160	2021
			Deepwater	5.6	9	4	0	58	2009
Regular *Oil* excludes oil from coal, shale, bitumen, heavy deepwater, polar & gasfield NGL			Polar	0.9	1	2	0	52	2030
			Gas Liquid	8.0	9	10	8	275	2027
			Rounding		2	-2		5	
Revised 26/12/2004			**ALL**	**83**	**85**	**65**	**35**	**2400**	2007

Figure 7b. Estimated world oil production to 2100. (Source: ASPO)

From the standpoint of the article's contribution to advancing the discussion beyond Hubbert's initial projections, its explanation of the methods and problems of estimating the world URR deserves treatment here. Many oil analysts have discounted warnings from Hubbert and his followers because official figures suggest that world oil reserves have grown substantially over the past 20 years. Campbell and Laherrère point out that such figures contain systematic errors arising from the fact that OPEC countries are often motivated to inflate reserve figures because the higher their reserves, the more oil they are allowed to export.

"There is thus good reason to suspect that when, during the late 1980s, six of the 11 OPEC nations increased their reserve figures by colossal amounts, ranging from 42 to 197 percent, they did so only to boost their export quotas," according to Campbell and Laherrère, who call such reserve growth "an illusion." They note that:

> about 80 percent of the oil produced today flows from fields that were found before 1973, and the great majority of them are declining.
>
> In the 1990s oil companies have discovered an average of seven Gbo [billion barrels of oil]; last year they drained three times that much. Yet official figures indicated that proved reserves did not fall by 16 Gbo, as one would expect; rather, they expanded by 11 Gbo. One reason is that several dozen governments opted not to report

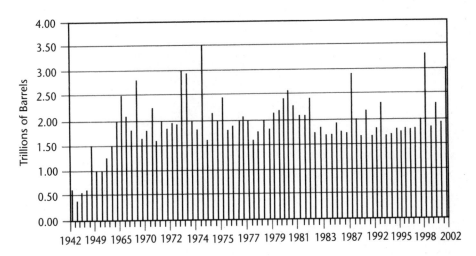

Figure 8. Published estimates of global ultimately recoverable oil, in trillions of barrels

(Source: C. J. Campbell)

declines in their reserves, perhaps to enhance their political cachet and their ability to obtain loans. A more important cause of the expansion lies in revisions: oil companies replaced earlier estimates of the reserves left in many fields with higher numbers. For most purposes, such amendments are harmless, but they seriously distort forecasts extrapolated from published reports.

Campbell and Laherrère suggest that one way to avoid such distortions is to backdate every revision to the year in which the field in question was first discovered. When this is done, it becomes apparent that global oil discovery peaked in the early 1960s and has been falling ever since. If that trend in discovery is extrapolated, it is possible to make a good guess at how much oil will ultimately be found. Even if this guess is off by two or three hundred billion barrels, the error will not affect the timing of the production peak by more than a few years.

The authors also discussed "nonconventional" oil — including heavy oil in Venezuela and oil sands in Canada — of which vast quantities are known to exist. "Theoretically," they write, "these unconventional oil reserves could quench the world's thirst for liquid fuels as conventional oil passes its prime. But the industry will be hard-pressed for the time and money needed to ramp up production of unconventional oil quickly enough." (Later in this chapter we will

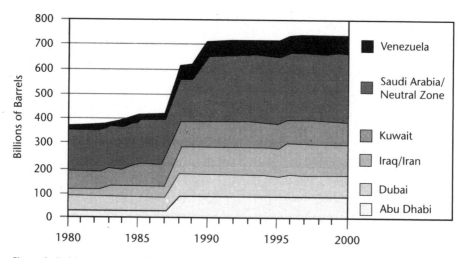

Figure 9. Dubious reserve revisions by OPEC countries in 1986 and 1987, in billions of barrels (Source: C. J. Campbell)

examine some other problems with nonconventional petroleum resources.)

Campbell is currently predicting the peak of global production of conventional oil to occur by about 2008.[18]

Kenneth S. Deffeyes (whom we quoted earlier in this chapter), in his book *Hubbert's Peak: The Impending World Oil Shortage* (2001), discusses the work of a petroleum geologist in layman's terms. The reader learns how oil was formed, where it is likely to be found, and what techniques and machinery geologists use to find it. Deffeyes also devotes two chapters to a detailed analysis of Hubbert's predictive method, offering mathematical refinements that yield more accurate forecasts. At the conclusion of the first of those chapters he writes:

> The resulting estimate gives a peak production year of 2003 and a total eventual oil recovery of 2.12 trillion barrels. The peak year, 2003, is the same year that we got by fitting [Colin] Campbell's 1.8-trillion-barrel estimate to the production history. Other published estimates, using variations on Hubbert's methods, give peak years from 2004 to 2009. I honestly do not have an opinion as to the exact date for two reasons: (1) the revisions of OPEC reserves may or may not reflect reality; (2) OPEC production capacities are closely guarded secrets .... This much is certain: no initiative put in place starting today can have a substantial effect on the peak production year. No Caspian Sea exploration, no drilling in the South China Sea, no SUV replacements, no renewable energy projects can be brought on at a sufficient rate to avoid a bidding war for the remaining oil. At least, let's hope that the war is waged with cash instead of with nuclear warheads.[19]

The late **L. F. Ivanhoe** was the founder of the M. King Hubbert Center for Petroleum Supply Studies at the Colorado School of Mines, whose mission is to assemble, study, and disseminate global petroleum supply data. He was a registered geologist, geophysicist, engineer, and oceanographer with 50 years of domestic and international experience in petroleum exploration with various private and government oil companies. He was associated first with Chevron and then with Occidental Petroleum, where he was senior advisor of worldwide evaluations of petroleum basins from 1974–80. Ivanhoe was the author of many papers on technical subjects, including roughly 50 on the evaluation of foreign prospective basins and the projection of future global oil supplies.

Ivanhoe called Hubbert's followers "Cassandras," after the mythological Trojan princess who could foretell the future but was doomed never to be believed.

In 1997, in a paper entitled "King Hubbert — Updated," Ivanhoe presented the following scenario:

> Hubbert wrote virtually nothing about details of the "decline side" of his Hubbert Curve, except to mention that the ultimate shape of the decline side would depend upon the facts and not on any assumptions or formulae. The decline side does not have to be symmetrical to the ascending side of the curve — it is just easier to draw it as such, but no rules apply. The ascending curve depends on the skill/luck of the explorationists while the descending side may fall off more rapidly due to the public's acquired taste for petroleum products — or more slowly due to government controls to reduce consumption .... [20]

In his summary at the end of that paper, Ivanhoe concluded that the

> critical date ... when global oil demand will exceed the world's production will fall somewhere between 2000–2010, and may occur very suddenly due to unpredictable political events .... This foreseeable energy crisis will affect everyone on earth.

Walter Youngquist, retired Professor of Geology at the University of Oregon, is the author of *Geodestinies: The Inevitable Control of Earth Resources over Nations and Individuals* (1997). During his career, he led or participated in on-the-ground geological studies in the US and abroad, and studied populations and resources in 70 countries. In his book, Youngquist discusses the important concept of net energy. He writes:

> All this energy expended in thousands of ways used to finally discover oil and produce it has to be added up and compared with the amount of energy in the oil which these efforts produce. This ratio — of energy produced compared to the energy used — is the all-important energy/ profit ratio. As we have to drill deeper to find oil, and as we have to move into more difficult and expensive areas in which to operate, the ratio of [energy] profit to energy expended declines. Already, in some situations energy in the oil found is not equal to the total energy expended. Also, although some wells flow

initially, all wells eventually have to be pumped. Pumping oil is expensive, particularly if it is being pumped from a considerable depth. It takes energy to move steel pumping rods up and down, in some cases as much as three miles of them ....

The most significant trend in the US oil industry has been the decline in the amount of energy recovered compared to energy

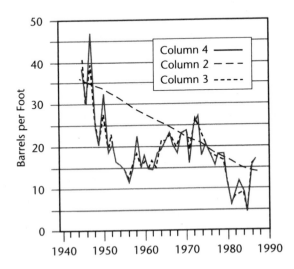

Figure 10. Yield per effort (YPE) for the US lower 48 states (onshore plus offshore). Yield per effort is the ratio of total annual additions to proved oil reserves to total oil footage drilled. The dark circles are actual observations of YPE. predicted by a regression that predicts YPE as a function of cumulative drilling effort alone. The solid line is YPE predicted by the regression that predicts YPE as a function of cumulative driling, the rate of drilling, revisions as a fraction of total additions, and new field discoveries as a fraction of the sum of new field discoveries, new discoveries in old fields, and extensions.
(Source: Oil Analytics)

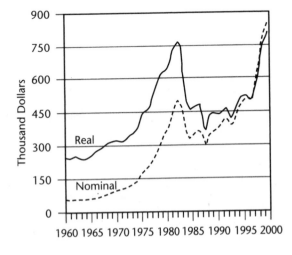

Figure 11. Costs per well of oil and gas wells drilled in the US, 1960–2000
(Source: US Energy Information Administration.)

expended. In 1916 the ratio was about 28 to 1, a very handsome energy return. By 1985, the ratio had dropped to 2 to 1, and is still dropping. The Complex Research Center at the University of New Hampshire made a study of this trend and concluded that, by 2005 at the latest, it will take more energy, on the average, in the United States to explore for, and drill for, and produce oil from the wells than the wells will produce in energy.[21]

About the end of oil production, Youngquist has this to say:

> Most likely the end of the Petroleum Interval will be gradual wherein no crisis point is reached, just slow change. But, especially with continually rising populations, and no sufficient substitutes for oil at hand, there is the possibility of a chaotic breakdown of society.[22]

Some of the essential elements of Hubbert's message have been taken up by others who are not petroleum geologists. A prominent example is **Matthew Simmons,** the founder of Simmons & Company International, an independent investment bank specializing in the energy industry. Simmons, has highlighted the reality and significance of petroleum depletion in many of his writings and public presentations.

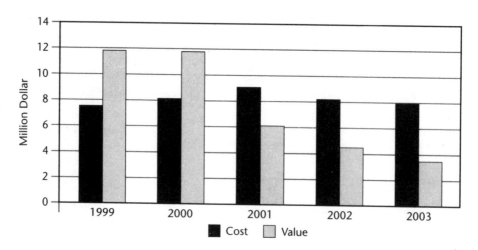

Figure 12. In recent years, the cost of exploration for oil has been exceeding the net present value of the discoveries in absolute terms. In simple terms, these days it usually costs more to explore for oil than consequent oil discoveries warrant. This trend appears to be accelerating. (Source: ASPO)

Simmons describes himself as a lifelong Republican with 30 years of experience in investment banking. In a lecture called "Digging Out of Our Energy Mess," delivered to the American Association of Petroleum Geologists on June 5, 2001, Simmons noted:

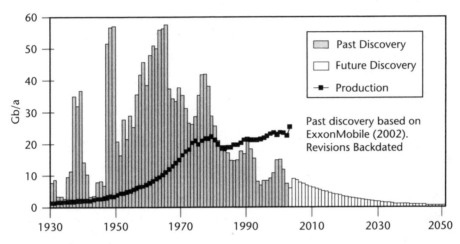

Figure 13a. The growing gap. (Source: ASPO)

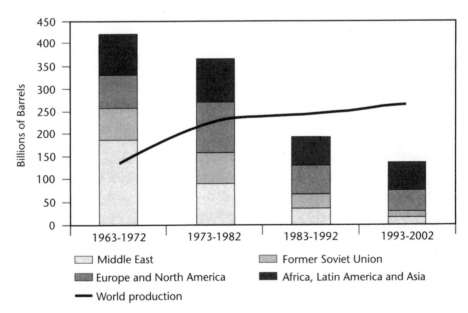

Figure 13b. Additions to world proven reserves from the discovery of new fields, and production.
(Source: World Energy Outlook 2004 [Figure 3.14])

A simple check of the facts quickly reveals that almost every scrap of spare energy [production] capacity around the globe is now either gone or just about to disappear ....

Even the Middle East is now beginning to experience, for the first time ever, how hard it is to grow production once giant fields roll over and begin to decline. There is so little data on field-by-field production statistics in the Middle East that any guesses on average decline rates are simply speculation. But there is growing evidence that almost every giant field in the Middle East has already passed its peak production.

It is also interesting to see how few truly giant oil and gas fields have been discovered over the past 40 years .... Even the newest giant field in the Middle East, Saudi's Shaybah field, was discovered in the 1970s, though it only began production two years ago.[23]

Simmons went on to point out that two of the world's other giant fields — Alaska's Prudhoe Bay and Western Siberia's Samatlor field, both discovered in 1967 — peaked in the 1970s, Prudhoe at 1.5 million barrels per day and Samatlor at over 3.5 million barrels per day:

Today, Prudhoe Bay struggles to stay around 500,000 barrels per day while Samatlor's production averaged just under 300,000 barrels per day in 2000. To think that two giant fields which collectively topped 5 million barrels per day 12 years ago could now be down to 800,000 barrels per day is a staggering example of the power of the decline curve ....

Could the days of 1 million barrel a day or greater oil and gas fields be over? There are over 140 oil and gas fields under development through the end of 2005. Only a handful of these projects have plans to produce over 200,000 barrels per day when they peak and none are expected to exceed 250,000 barrels per day.

Matt Simmons has recently authored a book, *Twilight in the Desert*, in which he analyzes over 200 technical papers by geologists working the Saudi fields, and concludes that Saudi Arabia's oil production may already be near or at its all-time peak. Since virtually all of the world's spare production capacity is in Saudi Arabia, that country's peaking date will also signal the global peak.

Zeroing in on the Date of the Peak

As a growing community of scientists applies itself to the task of determining exactly when the world's oil production will begin to decline, four principal methods for doing so are emerging.

1. Estimate the total ultimately recoverable resource (URR) and calculate when half will have been extracted. This is the original method developed and used by Hubbert himself, beginning in the 1950s. As we have seen, Hubbert noted that for a typical oil-producing region, a graph of extraction over time tends to take on a bell-shaped curve: the more cheaply and easily accessed portion of the resource is depleted first, so that when about half is gone the rate at which extraction can proceed tends to diminish.

The method is relatively simple, and it worked well for predicting the US production peak. However, it relies on the availability of accurate discovery and reserve data so that URR can be accurately estimated, and on accurate extraction data as well. For the US, these figures were not problematic, but for some other important oil-producing regions this is not the case: reserve data for Saudi Arabia, for example, are controversial.

Moreover, there is no natural law stating that the extraction curve must be precisely bell-shaped or symmetrical. Indeed, political, economic, and techno-logical factors can deform the curve in an infinite number of ways. In reality, the actual production curves from producing nations never conform to a sim-ple mathematical curve, and are characterized by bumps, plateaus, valleys, and peaks of differing sizes and durations. A war, a recession, the application of a new recovery technology, or the decision by a government to restrain extraction can reshape the curve arbitrarily.

Nevertheless, what comes up must come down: for every oil-producing region, extraction starts at zero, increases to a maximum, and declines, regard-less of the tortuous bumps in between. And, in general, the latter half of the resource will require more effort and time to extract than the first half. Moreover, experience shows that if actual production strays far from the predicted curve because of political or economic factors, it will tend to return to it once the influence of those factors subsides.

The Hubbert model is therefore a good way of providing a first-order approximation: it gives us a general overview of the depletion process, and (depending on the accuracy of the available reserve and production data) yields a likely peak year, with a window of uncertainty. However, it cannot be used to forecast actual production for the next month or even the next decade.

Using this method (with various refinements), Kenneth Deffeyes, in his book *Hubbert's Curve,* arrives at a peak date of 2005. Other researchers, such as Jean Laherrère, using more optimistic reserve estimates, place the peak further out, up to 2020.

2. Count the number of years from peak of discovery. Hubbert realized that, when graphed over time, the *discovery* of oil within any given region tends to peak and decline, just as production does. Understandably, discovery always peaks first — since it is necessary to find oil before it can be extracted.

In the US, discoveries of oil peaked in the early 1930s with the stupendous finds in east Texas; production peaked almost exactly 40 years later. Are we likely to find a similar time lag for other oil producing regions? If so, this could provide a basis for predicting the timing of the global production peak.

The duration of the lag between discovery and production peaks depends on a number of factors: the geological conditions (some fields can be depleted more quickly than others); the extraction technology being used (new recovery methods can deplete a reservoir more quickly and efficiently, but can also increase the total amount recoverable and thus extend the life of the field); and whether the resource is being extracted at maximum possible rate (as already noted, economic or political events can intervene to reduce production rates).

The North Sea provides an example of a relatively brief lag between discovery and extraction peaks: there, discoveries peaked in the early 1970s, while production peaked only 30 years later, at the turn of the new century. The latest exploration and extraction technologies were applied, and the resource base was drawn down at virtually the maximum possible rate because North Sea oil was in high demand throughout this period.

Iraq provides a counterexample: there, two principal periods of major discovery occurred — in the early 1950s and the mid-1970s. For that country, political and economic events have constrained production to a very significant degree: first, the Iran-Iraq war of the 1980s, then the US-led embargo of the 1990s, and finally the turmoil surrounding the US invasion and occupation have reduced extraction rates well below levels that would otherwise have been achieved. Consequently, Iraqi oil production may not peak until 2015 at the earliest, though more likely a decade or so later, yielding a discovery-to-production-peak lag of 45 to 60 years.

Will other discovery-to-production-peak lag times tend to more closely match those of the North Sea countries, or that of Iraq? Chances are that, as

The US Department of Energy Discusses Peak Oil

When *The Party's Over* was originally published, no official US or international agency had formally acknowledged the likelihood that global oil production will begin its historic decline within the next few years. That situation has recently changed. In March 2004, the Department of Energy published a little-heralded document on the strategic importance of oil shale; roughly a quarter of the 45-page report is devoted to the subject of oil depletion and its likely consequences. Here are just a few excerpts:

[World] Discoveries did peak before the 1970s, as shown in Figure 6. This figure also shows that no major new field discoveries have been made in decades. Presently, world oil reserves are being depleted three times as fast as they are being discovered . . . .

The disparity between increasing production and declining reserves can have only one outcome: a practical supply limit will be reached and future supply to meet conventional oil demand will not be available. The question is when peak production will occur and what will be its ramifications. Whether the peak occurs sooner or later is a matter of relative urgency . . . .

In spite of projections for growth of non-OPEC supply, it appears that non-OPEC and non-Former Soviet Union countries have peaked and are currently declining. The production cycle of countries . . . and the cumulative quantities produced reasonably follow Hubbert's model. . . . Although there is no agreement about the date that world oil production will peak, forecasts presented by USGS geologist Thomas Magoon, the OGJ [*Oil & Gas Journal*], and others expect the peak will occur between 2003 and 2020 . . . . What is notable about these predictions is that none extend beyond the year 2020 . . . . [pp. 7-8]

The Nation must start now to respond to peaking global oil production to offset adverse economic and national security impacts. [p. 26]

(Source: "Strategic Significance of America's Shale Oil Resource," Vol. 1, "Assessment of Strategic Issues," Office of Deputy Assistant Secretary for Petroleum Reserves, Office of Naval Petroleum and Oil Shale Reserves, U.S. Department of Energy, March 2004 <www.fe.doe.gov/programs/reserves/publications/Pubs-NPR/npr_strategic_significancev1.pdf>) ∎

individual anomalies cancel each other out, lag times are on average likely to cluster around that of the US — roughly 40 years.

Global oil discoveries peaked in 1963. This is not a controversial fact: both the oil industry and the US Department of Energy acknowledge that this is the case.

Given this, we might expect that the global peak in the rate of oil extraction would occur roughly 40 years later — i.e., in 2003.

However, we must take into account intervening economic and political events that might have tended to reduce extraction rates below their potential, and thus increase the time lag. The principal such events were the Arab OPEC embargo of the early 1970s, the fall of the Shah of Iran, and the subsequent Iran-Iraq war a few years later. The consequent oil price spikes reduced demand for oil, and led to a decline in extraction rates. The effect may have been to add up to ten years to the global discovery-to-production-peak time lag, yielding a likely peak date window of 2005 to 2013.

3. Track the reserve and production data of individual countries. For the past few years, both Colin Campbell (Association for the Study of Peak Oil) and Richard C. Duncan (Institute on Energy and Man) have been keeping close track of production data for individual producing nations.

Campbell's detailed discussions of oil statistics nation by nation are available in the archived newsletters of the Association for the Study of Peak Oil (www.asponews.org), and in his book, *The Essence of Oil & Gas Depletion* (MultiScience, 2003).

Duncan uses a "graphical-heuristic-iterative (GHI)" method to forecast world oil production, repeating the entire modeling and forecasting process annually to give a series of consistent but unique world oil forecasts. According to Duncan, *heuristic* means "a method of computer programming in which the modeler and machine proceed along empirical lines, using data, other information, and rules of thumb to find solutions or answers." *Iterative* means "repetitious; repeating or repeated." The *Graphical Input Device* (used in system dynamics programs such as Stella) enables the modeler "to quickly create and/or edit an oil production forecast of a nation just before each trial run (iteration) of the model." The *Scatter Graph* (system dynamics) is used to depict "the forecasted peak year of oil production (x-axis) *versus* the forecasted peak production rate (y-axis) of our ongoing series of world oil forecasts." Duncan describes this as a work in progress "that will eventually converge on Peak Oil — whether the Peak is near at hand or far in the future."[24]

Many countries are now clearly past their individual all-time extraction peaks. The list includes not only the US, but also Indonesia, Gabon, Great Britain, and Norway. Altogether, according to Duncan, of 45 significant oil-producing countries, 25 are past-peak (*BP Statistical Review of World Energy* currently estimates the latter number at 18, indicating that there is some uncertainty on this point, but also that oil companies are keenly aware of the peaking phenomenon and are keeping score).[25]

Some of the pre-peak nations are major producers with huge reserves (e.g., Iraq and Saudi Arabia). Thus it would be unwise to assume that the global peak will occur when exactly half of all producing nations have undergone their individual peaks. Clearly, more complex calculations are necessary, and this is the work that Duncan and Campbell are undertaking.

The countries in decline account for about 30 percent of the world's total oil production. Further, according to *Oil & Gas Journal,* as demand for oil expanded and prices rose during 2004, all of the added supply came from Russia and a few OPEC nations. Evidently, all of the nations outside of Russia and OPEC, when taken together, have already peaked in production (though there are individual exceptions, such as Brazil).

By examining the geology, history, and economic-political circumstances of each oil-producing country, it is possible to encircle the remaining uncertainties and pick away at them. How much oil has been discovered in each given nation? How long ago did discoveries peak? Are significant future discoveries likely? What kinds of recovery methods are being used?

Duncan summarizes his method as follows:

> I make a separate computer-based model to forecast the oil production for each of the major oil-producing nations in the world; 2) The latest oil data and related information on each nation are gathered from journals, the internet, and colleagues just before each national model is run; 3) Then both a Low oil forecast and a Medium oil forecast are made for each nation; 4) Next all of the Medium oil forecasts are combined (added up) to give the world oil forecast; 5) This process is repeated annually as soon as new oil data and related information become available; 6) A Scatter Graph indicates that the eight world oil forecasts that we've completed so far seem to be converging on Peak Oil in 2006 or 2007; 7) One new forecast (point) will be added to the Scatter Graph each year until Peak Oil is confirmed.

Campbell's analysis of likely future oil production by individual producing nations yields a global peak date of 2008.

4. Compare the amount of new production capacity likely to be available over the coming years with the amount of production capacity needed to offset decline rates from existing fields. The global oil industry needs to develop new production capacity yearly, in order to meet new demand and offset declines in production rates from individual wells and producing regions already past their all-time peaks. Currently, the world produces about 83 million barrels per day of all petroleum liquids combined (conventional oil plus oil from tar sands, natural gas liquids, and so on). The IEA estimates that in 2005 the world will need another 1.5 million barrels per day of new production capacity in order to meet new demand, plus another 4 mb/d to offset declines from existing fields — a total of about 5.5 mb/d. In 2006, a slightly greater new quantity

Many Little Peaks, One Big One

Richard Duncan, of the Institute on Energy and Man, has compiled the following forecasts on oil production peaks for 45 nations comprising seven regions (combined they accounted for more than 98 percent of the world's oil production, as of yearend 2003). The data are extracted from *Duncan's World Oil Forecast #8*, and (along with his unique method of world oil forecasting discussed previously) they are published for the first time here.

Note that forecast #8 includes oil production as defined by the *BP Statistical Review of World Energy,* June 2004, p. 6: "Includes crude oil, oil sands, NGLs (natural gas liquids — the liquid content of natural gas where this is recovered separately). Excludes liquid fuels from other sources such as coal derivatives."

Nation Peak Year

US 1970	Peru 1980
Canada 2007 (Includes oil sands.)	Trinidad & Tobago 1978
Mexico 2007	Venezuela 1970 (Includes heavy oils.)
NORTH AMERICA – 1985	SOUTH & CENTRAL AMERICA – 2015
Argentina 1998	Denmark 2007
Brazil 2008	Italy 2007
Colombia 1999	Norway 2001
Ecuador 2013	Romania 1976 ☞

will be needed, and in 2007, more still. In the five years from 2005 to 2010 a total of over 35 mb/d of new production capacity will need to come online. (These figures are agreed upon by both industry and various governmental agencies.) A substantial effort is necessary, to say the least.

But where will all this new production capacity come from?

In general, new production capacity arises from three sources: the discovery of new resources; the development of previously discovered resources (including reserve growth and infill drilling); or the development of unconventional resources (which sometimes depends on the invention and implementation of new technologies).

It takes time and investment to develop new production capacity. Thus it is possible — though no simple matter! — to gather the necessary data, analyze it, and project how much new production capacity is likely to emerge over the

UK 1999

EUROPE – 2000

FORMER SOVIET UNION – 1987

Iran 1974

Iraq 1979

Kuwait 1972

Oman 2001

Qatar 2007 (A large fraction is natural gas liquids.)

Saudi Arabia 1979 (Awaiting 2004 data; Saudi reserves are hotly debated.)

Syria 2003

UA Emirates 2007

Yemen 2014

MIDDLE EAST – 2008

Algeria 2006

Angola 2011

Cameroon 1985

Congo (Brazzaville) 1999

Egypt 1993

Equatorial Guinea 2011

Gabon 1996

Libya 1970

Nigeria 2008

Sudan 2010

Tunisia 1980

AFRICA – 2008

Australia 2000

Brunei 1979

China 2008

India 2007

Indonesia 1977

Malaysia 2006

Papua New Guinea 1993

Thailand 2006

Vietnam 2006

ASIA PACIFIC – 2007

WORLD PEAK 2007

(Source: Richard C. Duncan, 01/03/05) ∎

next five years, given current rates of investment, the available technology, and the discoveries in place. (Even if a huge new discovery were to be made next year, it would probably be impossible to bring the oil from it into production before 2010.) Chris Skrebowski, editor of *Petroleum Review*, has done just that in his 2004 report, "Oil Field Megaprojects," sponsored by the Oil Depletion Analysis Centre (ODAC).

Skrebowski compiles and regularly updates the details of planned major production projects, as reported by the oil companies. The list contains data on all announced fields with at least 500 million barrels of estimated reserves, and on projects with the claimed potential to produce 100,000 barrels a day or more.

Skrebowski and ODAC analyzed 68 production projects with announced start-up dates ranging from 2004 through 2010, and found that they are likely to add about 12.5 million barrels per day of new production capacity. In a press release, he stated: "This new production would almost certainly not be sufficient to offset diminishing supplies from existing sources and still meet growing global demand," and that "even with relatively low demand growth, our study indicates a seemingly unbridgeable supply-demand gap opening up after 2007."[27]

"It is disturbing to see such a dramatic fall-off of new project commitments after 2007, and not more than a handful of tentative projects into the next decade," Skrebowski said. "This could very well be a signal that world oil production is rapidly approaching its peak, as a growing number of analysts now forecast, especially in view of the diminishing prospects for major new oil discoveries."

At the end of the day, there are still uncertainties. Major new oil discoveries are always possible, though increasingly unlikely. Probably the greatest uncertainty with respect to the timing of the global oil production peak is future demand. If the global economy fares well, then demand will increase and the peak will come sooner; if the economy falters, then the peak will come later. If the world stumbles into a full-fledged depression, the peak could be delayed significantly, and the effects of the phenomenon could be masked by other events.

Nevertheless, as we have seen, the results of the possible forecasting methods tend to converge. We are within only a few years of the all-time global oil production peak. We are virtually at the summit now, with almost no time left for maneuvering before the event itself is upon us.

Hubbert's Critics: The Cornucopian Argument

If, as Hubbert and his followers have said, the future of oil production could spell disaster for industrial societies, then it is vital that we examine the geologists' claims from every possible angle to determine whether or not they are correct. Are there critics who dispute Hubbert, Campbell, et al., and are their critiques valid?

There is a school of thought, whose ideas are voiced mostly by economists, that says there is plenty of oil. In this section we will examine the arguments of three such "cornucopians": Peter Huber, Bjørn Lomborg, and Michael C. Lynch.

It is important to point out, however, that the cornucopian perspective is not limited to a few economists or industry lobbyists. As we will see, the USGS and Department of Energy (DoE) have posted petroleum production forecasts that are far more optimistic than those of Hubbert and his followers. These organizations present "official" projections, which are presumably supported by hard evidence.

Who is right? Sorting out the arguments is no small task, but the stakes are high enough to warrant whatever intellectual effort is required.

Let us begin with the most extravagant and general cornucopian claims, and work our way toward more specific and technical arguments.

Peter Huber, author of *Hard Green: Saving the Environment from the Environmentalists*, is a lawyer and writer. He earned his doctorate in mechanical engineering from MIT and served as an Assistant and later Associate Professor at MIT for six years. He is currently a senior fellow at the Manhattan Institute.

In an article entitled "The Energy Spiral" (2002), Huber claimed that the more energy humans use, the more they will be able to produce. According to Huber, hundreds of millions of years of biological evolution prove that nature is always finding ways of putting more energy to use. As manifestations of nature, human societies have likewise learned to obtain ever-greater amounts of energy; Huber calls this a "chain-reaction process," even a "perpetual-motion machine." In his view, the notion that humanity could ever run out of energy is absurd because the "more we capture and burn, the better we get at capturing still more."[28]

Huber appears to be telling us that the more cake we eat, the more we will have. This may be a cheerful message, but is it believable? True, living things have evolved to capture more and more energy from their environments. But we may be mistaken in conflating that biological capture of solar energy, whose growth trajectory leveled off hundreds of millions of years ago and may actually have peaked in the Mesozoic era, with human drawdown of fossil fuels, which began only centuries ago and is still spiraling upward at an astonishing rate. The latter process perhaps more closely resembles typical bloom-and-dieoff events, as when yeast cells are introduced into a wine vat filled with grape juice. With plenty of food available, the yeast organisms at first proliferate wildly. Their capture of the energy from their environment of sugar-laden juice grows exponentially — until their own fermentation byproducts begin to smother and poison them, whereupon all the organisms die.

Here is the essence of Huber's fallacy: he describes evolution as a one-way street — with species capturing ever-more energy — but omits any mention of the innumerable casualties that litter its curbs. For a species to run out of energy is hardly unprecedented; that's what extinction is all about, and vastly more species succumbed to extinction in the past than exist today. Moreover, as was discussed in Chapter 1, history is full of examples of complex human societies that overspent their energy budgets and collapsed as a result. There is no natural law that exempts modern industrial societies from the limiting principles that govern other living systems.

When we analyze it, Huber's argument amounts merely to a flawed and misapplied analogy.

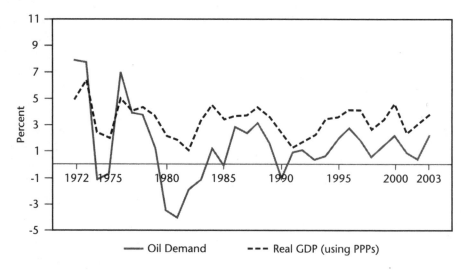

Figure 14. Relation between oil demand and GDP growth. Except during the 1970s and 1980s, when most of the world's nuclear power plants came into operation and reduced the demand for oil to fuel electricity generation, there is clearly a strong correlation. When many countries ceased adopting more nuclear power, oil demand accordingly grew by about 2 percent to deliver the ensuing growth of GDP.

(Source: International Energy Agency, "World Energy Outlook 2004.")

A somewhat more formidable critique of oil-depletion warnings is offered by **Bjørn Lomborg**, author of *The Skeptical Environmentalist* (2001) and Associate Professor of political science at the University of Aarhus, Denmark. In an article titled "Running on Empty" (2001), Lomborg writes:

> Today, oil is the most important and most valuable commodity of international trade, and its value to our civilisation is underlined by the recurrent worry that we are running out of it. In 1914, the US Bureau of Mines estimated that supplies would last only 10 more years. In 1939, the US Department of the Interior predicted that oil would last only 13 more years. In 1951, it made the same projection: oil had only 13 more years ....[29]

These predictions were obviously wrong. More recently, however,

> ... we have had an ever-rising prediction of the number of years' worth of oil remaining (years of consumption), despite increasing consumption. This is astounding. Common sense dictates that if we had 35 years' consumption left in 1955, we should have had 34 years' supply left the year after — if not less, because we consumed more oil in 1956 than in 1955. But ... in 1956 there were more years of reserves available ....

So how can we have used ever more, and still have ever more left? The answers provide three central arguments against the limited resources approach.

The first of Lomborg's "central arguments" is that "known reserves" are not finite but constantly growing:

> It is not that we know all the places with oil, and now just need to pump it up. We explore new areas and find new oil. It is rather odd that anyone could have thought that known resources pretty much represented what was left, and therefore predicted dire problems when these had run out. It is like glancing into my refrigerator and saying: "Oh, you've only got food for three days. In four days you will die of starvation." But in two days I will go to the supermarket and buy more food. The point is that oil will come not only from the sources we already know, but also from many sources of which we do not yet know.

His second argument is that we are constantly becoming better at exploiting resources:

We use new technology to extract more oil from known oilfields, become better at finding new oilfields, and can start exploiting oilfields that were previously too expensive and/or difficult to exploit. An initial drilling typically exploits only 20 percent of the oil in the reservoir. Even with the most advanced techniques using water, steam or chemical flooding to squeeze out extra oil, more than half the resource commonly remains in the ground. It is estimated that the 10 largest oilfields in the US will still contain 63 percent of their original oil when production closes down. Consequently, there is still much to be reaped in this area. According to the latest US Geological Survey assessment, such technical improvements are expected to increase the amount of available oil by 50 percent.

At the same time, we have become better at exploiting each litre of oil. Since 1973, the average US car has improved its mpg by 60 percent. Home heating in Europe and the US has improved by 24–43 percent. Many appliances have become much more efficient — dishwashers and washing machines have cut energy use by about 50 percent ....

Lomborg's third argument is that we can always find substitutes for any resource that begins to grow scarce:

We do not demand oil as such, but rather the services it can provide. Mostly we want heating, energy or fuel, and this we can obtain from other sources, if they prove to be better or cheaper. This happened in England around 1600 when wood became increasingly expensive (because of local deforestation and bad infrastructure), prompting a gradual switch to coal. During the latter part of the 19th century, a similar move from coal to oil took place.

In the short run, it would be most obvious to substitute oil with other commonly known fossil fuels such as gas and coal. For both, estimates of the number of years' supply remaining have increased. Moreover, shale oil could cover a large part of our longer-term oil needs. At $40 a barrel (less than one-third above the current world price of crude), shale oil can supply oil for the next 250 years at current consumption; in total, there is enough shale oil to cover our total energy consumption for 5,000 years.

In the long run, renewable energy sources could cover a large part of our needs. Today, they make up a vanishingly small part of global energy production, but this will probably change. The cost of solar energy and wind energy has dropped by 94–98 percent over the past 20 years, and they have come much closer to being strictly profitable. Renewable energy resources are almost incomprehensibly large. The sun could potentially provide about 7,000 times our own energy consumption — in principle, covering just 2.6 percent of the Sahara desert with solar cells could supply our entire needs.

It is likely that we will eventually change our energy uses from fossil fuels towards other, cheaper energy sources — maybe renewables, maybe fusion, maybe some as yet unthought-of technology. As Sheikh Yamani, Saudi Arabia's former oil minister and a founding architect of Opec, has pointed out: "The stone age came to an end not for a lack of stones, and the oil age will end, but not for a lack of oil." We stopped using stone because bronze and iron were superior materials; likewise, we will stop using oil when other energy technologies provide superior benefits.

I have quoted Lomborg at some length because he presents his ideas well and forcibly, and because the arguments he advances are the principal ones also cited by other Hubbert-school critics. Let us examine each of his points in turn, beginning with his preliminary comments.

The fact that some early oil-depletion predictions have failed does not tell us that all such predictions are bound to fail. Each prediction must be assessed on its own merits.

Moreover, the work of Hubbert and his followers is based on far better data and a far more robust understanding of the process of oil depletion than was available in the early 20th century. Hubbert predicted that US oil production would peak around 1970; it did. By now, roughly two dozen other oil-producing nations have passed their all-time production peaks. Nearly every year, another nation joins the "past-peak" club. Thus the discussion of the phenomenon of peak oil is as much about history as it is about prediction. The degree of extrapolation needed narrows with each passing year.

Why was there apparently more oil in the ground in 1956 than in 1955? Because these were some of the best years in history for oil discovery worldwide. Discovery rates have fallen off dramatically since then. The rate of discovery of new oil in the lower-48 US peaked in the 1930s; discovery worldwide peaked in the 1960s. Today, in a typical year, we are pumping and burning between five and six barrels of oil for each new barrel discovered. Demand for oil continues to increase, on average, at about two percent per year. From such information it should be possible to derive a working estimate of when global demand for oil will begin to exceed supply.

Now, to Lomborg's three main arguments. His first, that known reserves keep growing, centers on a subject to which Colin Campbell and Jean Laherrère have devoted years of study. As mentioned earlier, those authors have shown that such reserve growth is largely illusory and is derived partly from unverified and inflated reserve reports of OPEC countries vying for increased export quotas.

Lomborg implies that there is a vast amount of oil waiting to be discovered, but some specifics would be helpful. Where is all of this oil hiding? A few hints would surely cheer geologists who have spent decades applying the most advanced techniques to the problem of locating petroleum wherever it exists and who, on average, are finding ever smaller fields each year.

Lomborg's second argument is related to the first in that increased efficiency at recovering already discovered resources is often a component in the reported growth of existing oil reserves. Yes, new technology may enable us to increase the amount of oil extracted from any given field — perhaps, in some instances, even doubling the ultimately recoverable percentage. But enhanced recovery methods typically do not delay the peak of production from any given field by very much; they merely extend the field's production lifetime. Sometimes they merely enable recovery to proceed more rapidly, and thus cause the peak to

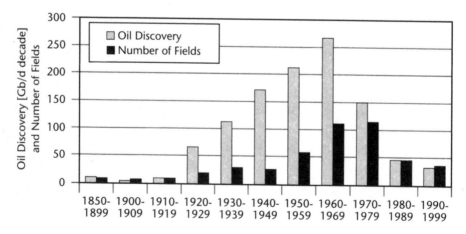

Figure 15. Giant oil field discoveries by decade.

(Source: International Energy Agency, "World Energy Outlook 2004.")

occur earlier. Campbell, Laherrère, *et al.*, have already accounted for such technology-based reserve growth in their estimates.

Moreover, it is important to understand that technology rarely offers a free ride; there are new costs incurred by nearly every technological advance. In the technologies involved with energy resource extraction, such costs are often reflected in the ratio of energy return on energy invested (EROEI). How much energy do we have to expend in order to obtain a given energy resource? In the early days of oil exploration, when we used simple technologies to access large, previously untapped reservoirs, the amount of energy that had to be invested in the enterprise was insignificant when compared with the amount harvested. As oil fields have aged and technologies have become more advanced and costly, that ratio has become less favorable.

This is reflected most clearly in figures for rates of oil recovery per foot of drilling. During the first 60 years of oil drilling (until 1920), roughly 240 barrels of oil were recovered, on average, for every foot of exploratory drilling. In the 1930s, as new geophysical exploratory techniques became available and the 6 billion-barrel east-Texas field was found by accident, the discovery rate reached a peak of 300 barrels per foot. But since then, during successive decades of drilling, discoveries per foot of drilling have dropped steadily to fewer than 10 barrels

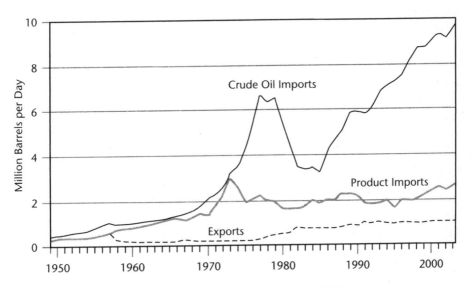

Figure 16. US trade in petroleum and petroleum products, 1949-2003

(Source: US Energy Information Administration)

per foot. And this decline has occurred during a period of intensive exploration, using ever more advanced technologies, such as 3D seismic and horizontal drilling. Thus, while new technologies have enabled the discovery of more oil, the EROEI for the activity of oil exploration has inexorably plummeted.

The same will no doubt be true of technologies used to increase the amount recoverable from existing reservoirs: we will indeed be able to get more oil out of wells than we otherwise would have, but we will have to invest more effort —and thus more energy — to obtain that oil, with an ever-decreasing EROEI.

How important is EROEI? When the EROEI ratio for oil exploration declines to the point that it merely breaks even — that is, when the energy equivalent of a barrel of oil must be invested in order to obtain a barrel of oil — the exercise will become almost pointless. Even if oil remains a useful lubricant or a feedstock for plastics, it will have ceased to be an energy resource. EROEI is also an essential consideration in the substitution of one energy resource for another: if we replace an energy resource that has, say, a four-to-one EROEI ratio with an alternative that has a two-to-one EROEI ratio, we will have to produce roughly twice as much gross energy to obtain the same net quantity. Thus, when a society adopts lower-EROEI energy sources, the amount of energy available to do work in that society will inevitably decline.[30]

The other half of Lomborg's efficiency argument is that we are learning to use each barrel of oil more thoroughly, thus getting more work out of it. This is certainly true and commendable, but it is a fact that must be viewed in context. The all-important context, in this instance, is that our total petroleum usage, nationally and globally, continues to increase each year. In terms of depletion rates and production peaks, *increased efficiency of use means nothing unless we are actually reducing the total amount of petroleum extracted and burned.* That is not happening, nor does any responsible agency project it to happen voluntarily within the next two or three decades. We are not reducing our dependency on oil — it is still growing.

Lomborg's third argument — that we can always find a substitute for any scarce resource — raises questions that we will address in more detail in the next chapter, in a discussion of alternative energy sources. For now, suffice it to say that substitutes, to be successful, must pass certain tests. When Europeans began substituting abundant coal for scarce wood, they soon found that their substitute sometimes contained more energy per kilogram than the original resource. When industrial countries began switching from coal to oil, the substitute was very noticeably more energy-dense. Lomborg suggests that industrial societies will deal with petroleum shortages by switching back to coal, but that means returning to a resource that is substantially *less* energy-dense and thus unsuitable for supplying society's vastly increased energy needs. He also mentions natural gas — but is there enough available to substitute for oil? Again, we will address that important question in detail in the next chapter; for now, it is enough merely to point out that North American production of natural gas is already in sharp decline.

Ah, but there is enough shale oil to last 5,000 years! Lomborg helpfully informs us that the dollar price of shale oil will necessarily be higher than the current price of conventional oil, which suggests a lower EROEI, but he does not discuss net-energy figures explicitly. Had he done so, the picture would not have been so encouraging.

Shale oil (or oil shale) is actually a misnomer: the rock is not shale but organic marlstone, and it contains no oil, but rather a solid organic material called *kerogen*. However, promoters have always preferred terms like "oil shale," since they encourage the sale of venture shares. Efforts to develop an oil-shale industry date back nearly 90 years, and so far all attempts — even serious and relatively recent ones by Chevron, Unocal, Exxon, and Occidental Petroleum — have failed. The recovery process involves mining ore, transporting it, heating it to 900 degrees Fahrenheit, adding hydrogen, and disposing of the waste — which is much greater in volume than the original ore and is also a groundwater pollution hazard. Processing and auxiliary support facilities require large amounts of fresh water — a resource intrinsically even more precious than oil. Walter Youngquist sums up the situation well: "Adding up the water supply problem, the enormous scale of the mining which would be needed, the low, at best, net energy return, and the huge waste disposal problem, it is evident that oil shale is unlikely to yield any very significant amount of oil, as compared with the huge amounts of conventional oil now being used."[31]

Lomborg might also have mentioned tar sands (sometimes optimistically called "oil sands"), which are likewise reputed to be potential substitutes for conventional oil. The Athabasca tar sands in northern Alberta are estimated to contain an estimated 870 billion to 1.3 trillion barrels of oil (when processed) — an amount equal to or greater than all of the conventional oil extracted to date. Currently, Syncrude (a consortium of companies) and Suncor (a division of Sun Oil Company) operate oil-sands plants in Alberta. Total production from the tar sands now stands at about one million barrels per day. The extraction process involves using hot-water flotation to remove a thin coating of bitumen from grains of sand, then adding naphtha — a petroleum distillate — to the resulting tar-like material in order to upgrade it to a synthetic crude that can be pumped. Currently, two tons of sand must be mined in order to yield one barrel of oil. As with oil shale, the net-energy figures for tar sands are discouraging: Youngquist notes that "it takes the equivalent of two out of each three barrels of oil recovered to pay for all the energy and other costs involved in getting the oil from the oil sands."[32]

The primary method that is used to process tar sands yields an oily waste water. For each barrel of oil recovered, two-and-a-half barrels of liquid waste are pumped into huge ponds. In the Syncrude pond, measuring 22 kilometers (14 miles) in circumference, six meters (20 feet) of murky water float on a 40-meter-thick (133-foot) slurry of sand, silt, clay, and unrecovered oil.[33] Residents of northern Alberta have initiated lawsuits and engaged in activist campaigns to close down the tar-sands plants because of devastating environmental problems associated with their operation, including the displacement of native peoples, the destruction of boreal forests, livestock deaths, and a worrisome increase in human miscarriages.[34]

To replace the global usage of conventional crude — 70 million barrels a day — would require about 350 additional plants the size of the existing Syncrude plant. Together, they would generate a waste pond of 8,750 sq. km, about half the size of Lake Ontario. But since tar sands yield less than half the net energy of conventional oil, the world would need more than 700 plants to supply its needs, and a pond of over 17,500 sq. km — almost as big as Lake Ontario. Realistically, while tar sands represent a significant energy asset for Canada, it would be foolish to assume that they can make up for the inevitable decline in the global production of conventional oil.

When examined closely, Bjørn Lomborg's arguments amount to an appeal to unspecified future discoveries and to hopeful but vague promises.

Michael C. Lynch, Chief Energy Economist of DRI-WEFA, Inc., has written extensively on petroleum depletion and is probably the foremost oil cornucopian in the current public debate. In his many writings he has emphasized essentially the same points as Lomborg, which we need not address again. However, in his essay "Closed Coffin: Ending the Debate on 'The End of Cheap Oil'" (2001), Lynch offers a confrontational, if somewhat technical, challenge leveled specifically at Campbell and Laherrère.[35] In it, he leaves aside other arguments and focuses almost entirely on reserve growth. I apologize to readers who are uninterested in this level of detail, but since the question of whether oil production is about to peak is central to this book, it is absolutely necessary that we examine the contentions of this foremost critic of production-peak estimates. Lynch writes:

> The primary flaw in [Campbell and Laherrère's] model is the assumption that recoverable petroleum resources are fixed, when the amount of oil which can be recovered depends on both the total amount of oil (a geological factor which is fixed), but also dynamic variables like price, infrastructure, and technology. If the amount of recoverable oil increases, as it has in the past, then the level predicted for peak production must increase and the date [of the production peak be] pushed further into the future ....
>
> The reliance on discovery trends to estimate URR has received similar criticism as the faulty URR estimates, namely that estimates of field size tend to increase over time with improved recovery methods, better examination of seismic data, infill drilling, and so forth. This means that the size of the recent fields is being underestimated compared to older fields .... An analogy would be to plant trees over twenty years and note that the size of the most recently planted trees was shrinking, and concluding that timber resources would become scarce ....

Following these general comments, Lynch makes his specific charges:

> Last year, the publication of the USGS's World Petroleum Assessment provided one particularly sharp nail in the coffin of this argument, when (among other things) they examined the development of field size estimates over time using the same proprietary database which Campbell and Laherrère relied on, and concluded that reserve growth from existing fields, although uncertain, would be substantial. They published a mean estimate of 612 billion barrels (nearly 30 years of current consumption) ....
>
> But the final nails seem to be located in this summer's little-noticed announcement by IHS Energy — the firm whose field database Campbell and Laherrère have utilized — of estimated discoveries. According to the firm, discoveries in 2000 were 14.3 billion barrels, a 10 percent drop from 1999. This has two interesting implications: first, discoveries have risen sharply the past two years, refuting the statement that poor geology, rather than lack of

access to the most prospective areas in OPEC, has kept discoveries low for the past three decades .... Undoubtedly [Campbell and Laherrère] — and others — will argue that this is due to the firm's inclusion of deepwater reserves, which they are not considering, and that is a factor in the recent robustness of discoveries. However, the primary element behind the greater discovery rates has been the finding of two new supergiant fields in Kazakhstan and Iran. Again, this refutes the argument that discoveries have been relatively low in recent decades due to geological scarcity and supports the optimists' arguments that the lower discoveries are partly due to reduced drilling in the Middle East after the 1970s nationalizations ....

[W]hile we need be concerned about quite a number of issues related to petroleum supply — depletion, change in reserve growth, concentration of production in politically stable areas — a possible near-term peak in production (conventional or otherwise) is not one of them.

As we did with Lomborg's arguments, let us address Lynch's one by one. His first substantive point has to do with the USGS "World Petroleum Assessment 2000," which predicts such substantial reserve growth as to delay a production peak by many years, perhaps by two or more decades.[36] The USGS is a government agency that employs many competent geologists and data analysts. Is there any reason to disbelieve its projections?

Many of the USGS's own experts criticize what they view as wildly optimistic assumptions contained in the WPA 2000 report. USGS geologist L. B. Magoon maintains a website warning of the imminent "Big Rollover" world production peak.[36]

In fact, the report's main authors, Schmoker and Klett, explain clearly in their chapter on reserve growth that there is complete uncertainty about reserve growth outside of the US and Canada, but that they believe it is better to use the US lower-48 reserve growth function than none at all. However, there are serious problems with extrapolating historic US reserve growth figures to the rest of the world.

The following example may be helpful. Assume a Texas oil field discovered in the 1930s. Examine the reported reserve growth from 1965 to 1995 — namely 30-year reserve growth figures for a 30-year-old field. Now apply this growth factor to a Saudi field discovered in 1965, using reported production and reserve figures as of 1995. The result: considerable growth is to be expected from the Saudi field. But there are two main problems with this method:

First, the 1930s Texas reserve estimate was probably intentionally under-stated, and the 1995 Saudi report was probably intentionally overstated. Typically, US oil companies have reported reserves with an extremely conservative 90 percent probability of recovery (P90), while other countries, including Saudi Arabia, use a 50 percent probability (P50) for their reserve estimates. Some countries even report P10 reserves, yielding greatly inflated figures. And, as we have already seen, the Saudis stated a substantial "proven" reserves addition in the late 1980s that was probably mostly, if not entirely, spurious. True Saudi reserve figures remain a state secret.

Second, US reserve growth after 1965 benefited from recent technological recovery advances and included the reporting of at least part of the previously understated reserves. The Saudi estimates of 1995, in contrast, already included the expected impact of all recent technological recovery advances, which became standard in the industry from the 1970s on.

Thus it is unreasonable to assume that the Saudi field will experience the same rate of reserve growth in the next three decades as the Texas field did in the past. If the USGS estimates were corrected for these problems, it is doubtful that what the authors call "potential" reserve growth would exceed 300 billion barrels, an amount that would not significantly affect projections for the peak production year.

But the USGS analysis is far more sanguine than this; it calls for a total increase of 1200 billion barrels of oil (discovery plus reserve growth) during the decades from 2000 to 2030, or an average increase of 40 billion barrels per year. During the most productive decade of discovery in world history — from 1957 to 1967 — exploration yielded an average of 48 billion barrels per year. If the industry is capable of repeating that feat, why hasn't it done so in any of the past three decades? Discovery plus reserve growth averaged 9 billion bar-rels per year in the decade of the 1990s. It is difficult to imagine circumstances that would enable that figure to quadruple in the years ahead.

Why would a government agency like the USGS publish a report that gives an extravagantly optimistic view of global oil resources? Nor is such optimism confined to the USGS: the Energy Information Agency (EIA) of the Department of Energy (DoE) has released similarly rosy projections. What's going on here?

A clue is contained in a sentence buried in the EIA "Annual Energy Outlook 1998 with Projections to 2020"; it reads: "These adjustments to the

USGS and MMS [Materials Management Service] estimates are based on non-technical considerations that support domestic supply growth to the levels necessary to meet projected demand levels."[37] In other words, supply projections were simply engineered to fit demand projections. As industry insiders have known for years, USGS and EIA data on current and past production are accurate as can be hoped for, given the fuzziness of the numbers from some producing countries. But their future projections are essentially political statements designed to convey the message that there is no foreseeable problem with petroleum supply and that the American people should continue buying and consuming with no care for the future. This is not a new situation: in 1973, Congress demanded an investigation of the USGS for its failure to foresee the 1970 US oil production peak.[38]

In contrast, the Paris-based International Energy Agency (IEA) adopted a modified Hubbert-peak forecasting method in 1998, predicting a production peak in 2015. Its *World Energy Outlook: 2001* concluded that soon all non-Middle East oil reservoirs will peak and decline, throwing the world into increasing dependence on a small number of Middle East suppliers.[39]

Next let us examine Lynch's claims concerning the world oil discovery figures for the years 1999–2000. They were, as he points out, anomalously large. Still, the amount discovered in the better year of the two — 1999 — represented only about 62 percent of the amount of all oil extracted and consumed that year. If, in even the best recent year of discovery, the world still used much more oil than was found, this is hardly an argument against the idea that production will foreseeably peak.

But let's look closer. The average figure for discovery plus reserve growth in the years 1996–2000 was about 10 billion barrels per year. Assuming that this rate could continue, we might, in the next 30 years, expect that 300 billion barrels of oil would be added to current proven reserves (let's use the credible estimate that 1100 billion barrels remain to be produced globally out of an original URR of 2000). Meanwhile, we must subtract the yearly projected drawdown of those reserves; with a conservatively estimated average demand of 30 billion barrels per year in the decades 2000–2030, that would be 900 billion barrels total. A quick calculation shows that half the oil would be gone — and hence production would likely peak — well before 2010. But remember: these figures are optimistic in every respect; we are assuming, for example, that many more large discoveries like the Kazakhstan find of 1999 will continue to occur, when the actual long-term trend is toward the discovery of less oil with each passing year.

Lynch believes that increased drilling in the Middle East and in deep-water areas will make all the difference. While more discovery will no doubt take place in the Middle East, most of the largest fields there were found in the 1960s. Nearly the entire region has been mapped with 3D seismic; and, due to the time interval needed to ramp up production, even the discovery tomorrow of a couple of more "elephants" in the range of 50 billion barrels each would not push back the global production peak by more than a few years. Deep-water reserves are challenging and costly to access — in both monetary and energy terms. And again, a few moderate-to-large discoveries in deep-water regions made now will not significantly delay the global production peak.

The following paragraph from Campbell and Laherrère's "The End of Cheap Oil" (1998) puts matters in perspective:

> Perhaps surprisingly, that prediction [of a production peak during the first decade of the new century] does not shift much even if our estimates are a few hundred billion barrels high or low. Craig Bond Hatfield of the University of Toledo, for example, has conducted his own analysis based on a 1991 estimate by the U.S. Geological Survey of 1,550 Gbo remaining — 55 percent higher than our figure. Yet he similarly concludes that the world will hit maximum oil production within the next 15 years. John D. Edwards of the University of Colorado published last August one of the most optimistic recent estimates of oil remaining: 2,036 Gbo. (Edwards concedes that the industry has only·a 5 percent chance of attaining that very high goal.) Even so, his calculations suggest that conventional oil will top out in 2020.

Tellingly, Michael Lynch refuses to offer his own prediction of when global oil production will peak, even when pressed to do so.

Who Is Right? Why Does It Matter?

In many two-sided controversies, the bystander is justified in assuming that both sides have valid points and that the truth probably lies roughly equidistant between extreme claims. But on the vital question of when world oil production will peak, the arguments of cornucopians like Huber, Lomborg, and Lynch appear vague and weak, and the assessments of public agencies like the USGS and EIA sometimes break down under close scrutiny. In contrast, the clarity and logic of the analysis, and the depth of expertise, of the petroleum pessimists — Campbell, Laherrère, Deffeyes, Youngquist, *et al.* — seem impressive.

Ultimately, we will know for sure when global oil production peaks only after the fact: one year we will notice that gasoline prices have been climbing at a rapid pace, and we will look back on the previous few years' petroleum production figures and note a downward slope. It is possible that the next decade will be a "plateau" period, in which recurring economic recessions will result in lowered energy demand, which will in turn temporarily mask the underlying depletion trend.

As I have made clear, I personally am convinced of the correctness of the Cassandras' message that global conventional oil production will peak some time during this first decade of the 21st century.

The world reached a fork in the road in the 1970s. In some respects it is still hesitating at that juncture. The two conflicting paths of action with which we were — and still are — presented correspond fairly closely with the "two universal, overlapping, and incompatible intellectual systems" mentioned by M. King Hubbert in the passage quoted earlier in this chapter.

On the one hand is the path based on the "monetary culture that has evolved from folkways of prehistoric origin." This is the path of the optimists, who are predominantly economists by profession (Michael Lynch is the prime example, though Peter Huber, who has an engineering degree, represents a counterexample). For decades most economists have been united in proclaiming that resources are effectively infinite, and that the more of any resource we consume, the more its reserves will grow. The human intellect is the greatest resource of all, the optimists tell us, and so population growth means that we all benefit from an increasing collective problem-solving capacity. Like money in the bank expanding inexorably through compounding interest, humanity is growing a measurably brighter future with each passing year as it reproduces, transforms its environment, invents new technologies, and consumes resources.

On the other hand is the path based on "the accumulated knowledge of the ... properties and interrelationships of matter and energy." For decades we have also been hearing from ecologists, petroleum geologists, climatologists, and other scientists who tell us that resources are limited, that the Earth's carrying capacity for humans is finite, and that the biosphere on which we depend cannot for long continue to absorb the rapidly expanding stream of wastes from industrial civilization.

Our leaders' hesitancy to listen seriously to the latter point of view is understandable; if they did so, they would logically and morally be compelled to

1. adopt the ethic of "sustainability" in all aspects of planning, thinking ahead for many future generations;

2. institute systematic efforts to improve efficiency in the use of energy, and combine such efforts with programs to reduce the total amount of energy used by society;

3. encourage the rapid development and deployment of all varieties of renewable energy technologies throughout society;

4. systematically discourage (through taxation or other means) the consumption of nonrenewable resources; and

5. find humane ways to encourage a reduction in human fertility in all countries, so as to reduce the population over time.

As a result of their inaction along these lines, our leaders have in effect chosen the first path, that of the optimists, which implies a diametrically opposite pattern of choices and compels them to

1. make plans to meet only short-term crises because that is the only kind we will ever face, and don't worry about future generations because they will have advanced technologies to solve whatever problems we may be creating for them;

2. forget about efforts to impose improvements in energy efficiency since the marketplace will provide for improvements when and if they are needed;

3. forget about government programs to develop renewable energies because if and when alternatives are needed, price signals will trigger the market to turn in their direction;

4. continue to use fossil fuels at whatever rates are dictated by the market since to do otherwise will hurt the economy; and

5. treat population growth as a benefit rather than a problem, and do nothing to slow or reverse existing growth trends.

This latter path involves less short-term intervention in the economy and works to the near-term advantage of many significant power holders in society (including the oil and automobile companies). By taking it, our politicians have simply followed the path of least resistance.

This may be understandable, but the consequences — if the economists are wrong and the physical scientists are right — will be devastating for nearly everyone.

It is therefore particularly important that we think long and hard about the path not taken before it disappears from sight altogether. What if the Cassandras are right?

Throughout the rest of this book — primarily because of what I see as the overwhelming hard evidence in its favor, but also for the reason just cited — I will assume as correct the Cassandras' prediction that global oil production (all liquids) will peak some time during the remainder of this decade.

If we take that as a given, can we still avoid catastrophe by switching to other technologies and fuels in the years ahead? What, precisely, are our options?

Non-Petroleum Energy Sources:
Can the Party Continue?

Under the rule of the "free market" ideology, we have gone through two decades of an energy crisis without an effective energy policy .... We have no adequate policy for the development or use of other, less harmful forms of energy. We have no adequate system of public transportation.

— Wendell Berry (1992)

The pattern of preferences for using energy efficiency to decrease demand and [for renewable energy sources] to supply energy has been consistent in the poll data for 18 years. This is one of the strongest patterns identified in the entire data set on energy and the environment.

— Dr. Barbara Farhar (2000)

Nonrenewable resources should be exploited, but at a rate equal to the creation of renewable substitutes.

— Herman Daly (1992)

Continuing to increase our dependency on petroleum consumption is clearly a suicidal course of action. The only intelligent alternative is to begin reducing energy consumption and finding alternative energy sources to substitute for petroleum.

— Paul Ehrlich (1974)

Total energy consumption is projected to increase from 96.1 quadrillion British thermal units (BTU) to 127.0 quadrillion BTU between 1999 and 2020, an average annual increase of 1.3 percent.

— US Department of Energy (1999)

137

T his chapter focuses exclusively on a single vital question: *To what degree can any given non-petroleum energy source, or combination of sources, enable industrial civilization to survive the end of oil?*

Before we can make this assessment, it is important that we clearly understand what has made oil such a valuable energy commodity. Oil is

- easily transported (liquid fuels are more economically transported than solids, such as coal, or gases, such as methane, and can be carried in ships far more easily than can gases);
- energy-dense (gasoline contains roughly 40 kilowatt-hours per gallon);
- capable of being refined into several fuels, including gasoline, kerosene, and diesel, suitable for a variety of applications; and
- suitable for a variety of uses, including transportation, heating, and the production of agricultural chemicals and other materials.

Moreover, historically petroleum has been easy to access, which has helped give it a very high energy return on energy invested (EROEI). Net energy — or EROEI — is a subject we will touch on frequently in this chapter. In assessing each of the non-petroleum energy sources, I will refer to net-energy figures from Howard T. Odum's *Environmental Accounting, Energy and Decision Making* (1996), and C. J. Cleveland, R. Costanza, C. A. S. Hall, and R. Kaufmann's "Energy and the U.S. Economy: A Biophysical Perspective" (1984).[1] Odum assigns imported oil a current EROEI of between 8.4 (that is, 8.4 units of energy returned on every unit of energy invested in exploration, drilling, building of drill rigs, transportation, the housing of production workers, etc.) and 11.1, depending on the source.

However, for the period between 1950 and 1970, he calculates that oil had an EROEI of 40. Cleveland *et al.* calculated a greater than 100-to-1 return for oil discoveries prior to 1950, which declined to a 30-to-1 return by the 1970s.

In this chapter we will examine each of the most prominent non-petroleum energy sources, starting with those that are closest to oil in their characteristics (i.e., the other fossil fuels: natural gas and coal), then moving to nuclear and geothermal power, the renewables (solar power, wind, biomass, tides, waves, and hydro), hydrogen, and exotic sources (cold fusion and "zero-point" energy). Finally, we will explore the potential for energy conservation (not a "source," but an essential strategy) to ensure the survival of industrial societies as the petroleum interval comes to a close.

Natural Gas

In some respects, natural gas appears to be an ideal replacement fuel for oil: it burns more cleanly (though it still produces CO_2); automobiles, trucks, and buses can be converted to run on it; and it is energy-dense and versatile. Its EROEI is quite high. It has long been used to create nitrogen fertilizers for agriculture (through the Haber-Bosch process), for industrial processes like glassmaking, for electricity generation, and for household cooking and heating. Currently, natural gas accounts for about 25 percent of US energy consumption; 17 percent of the gas extracted is used to generate electricity. Thus there already is an infrastructure in place to make use of this fuel.

Could extraction be increased to make up for the projected shortfalls in oil? Some organizations and individuals claim there is enough gas available globally to last for many decades. Estimates for total reserves vary from about 300 to 1,400 tcf (trillion cubic feet). With such a wide range of figures, it is clear that methods of reporting and estimating are imprecise and speculative. The number 1,100 tcf is often cited; this would represent 50 years' worth of reserves at current rates of global usage. The ever-optimistic US Energy Information Agency (EIA) reports that the US also has about 50 years' worth of natural gas, with proven reserves of 177.4 tcf in 2001. As of 2001, annual usage was in the range of 23 tcf.[2]

Clearly, the EIA is assuming considerable future discovery, as current proven reserves would last fewer than ten years at current usage rates. That assumption — that future discoveries will more than quadruple current proven reserves — is highly questionable; moreover, we should also ask: Does natural gas depletion follow a Hubbert-type curve, so that we should expect a peak of production and a long period of decline to occur long before the last cubic foot is extracted?

Many industry analysts believe the outlook for future discoveries in North America is far less favorable than EIA forecasts suggest. In the decade from 1977 to 1987, 9,000 new gas fields were discovered, but the following decade yielded only 2,500 new fields. This general downward trend in discovery is continuing, despite strenuous efforts on the part of the industry. Matthew Simmons has reported that the number of drilling rigs in the Gulf of Mexico grew by 40 percent between April 1996 and April 2000, yet production remained virtually flat. That is largely because the newer fields tend to be smaller; moreover, because of the application of new technology, they tend to be depleted faster than was the case only a decade or two ago: new wells average a 56 percent depletion rate *in the first year of production.*

In a story dated August 7, 2001, Associated Press business writer Brad Foss noted that in the previous year, "there were 16,000 new gas wells drilled, up nearly 60 percent from 10,400 drilled in 1999. But output only rose about 2 percent over the same period, according to estimates from the Energy Department. The industry is on pace to add 24,000 wells by the end of the year, with only a marginal uptick expected in production."[3]

In June 1999, *Oil & Gas Journal* described how the Texas gas industry, which produces one-third of the nation's gas, had to drill 6,400 new wells that year to keep production from plummeting. Just the previous year, only 4,000 wells had to be drilled to keep production steady.[4]

According to Randy Udall of the Community Office for Resource Efficiency in Aspen, Colorado, "[n]o one likes talking about [natural-gas] depletion; it is the crazy aunt in the attic, the emperor without clothes, the wolf at the door. But the truth is that drillers in Texas are chained to a treadmill, and they must run faster and faster each year to keep up."[5]

US natural gas production has been wavering for years; in order to make up for increasing shortfalls, the nation has had to increase its imports from Canada, and Canada is itself having to drill an increasing number of wells each year just to keep production steady — a sign of a downward trend in discovery. A May 31, 2002 article by Jeffrey Jones for Reuters, entitled "Canada Faces Struggle

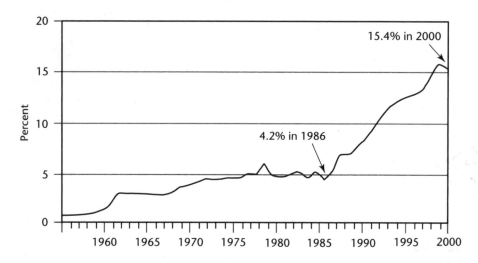

Figure 17. Net US imports of natural gas as share of consumption

(Source: US Energy Information Administration)

Pumping More Natgas to US," begins ominously: "Canadian natural gas production may have reached a plateau just as the country's role as supplier to the United States is becoming more crucial due to declining US gas output and rising demand ...."

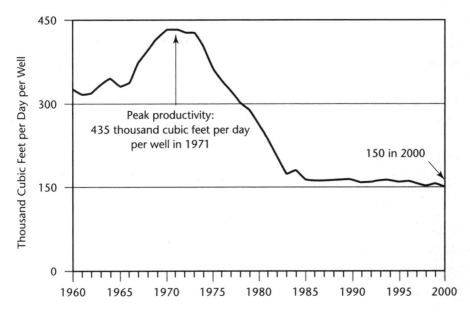

Figure 18a. US natural gas well productivity (Source: US Energy Information Administration)

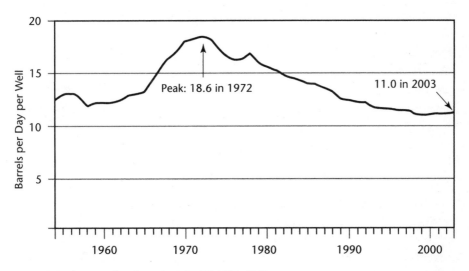

Figure 18b. Average oil well productivity, US 1950–2003 (Source: US Energy Information Administration)

Furthermore, Mexico has already cut its gas exports to the US to zero, and has become a net importer of the fuel.

A gas pipeline from Alaska could help, but not much. A three-foot-diameter pipeline would deliver only two percent of the projected needs for the year 2020.

Nearly all of the natural gas used in the US is extracted in North America. While gas is more abundant in the Middle East, which has over a third of the world's reserves, gas is not easily transported by ship. It must be cooled to minus 260 degrees Fahrenheit (minus 176 degrees Celsius) during the journey, requiring special liquid-natural-gas (LNG) tankers and ports. There are currently only three such ports in the US, though many more are being proposed.

Moreover, nearly all of the existing LNG shipping capacity is spoken for by Japan, Korea, and Taiwan through long-term contracts. Europe and the Far East may be able to depend on gas from the Middle East and Russia for several decades to come, but that is probably not a realistic prospect for the US.

The public got its first hint of a natural gas supply problem in the latter months of 2000, when the wellhead price shot up by 400 percent. This was a more dramatic energy price increase than even the oil spikes of the 1970s. Homeowners, businesses, and industry all suffered. This gas crisis, together with simultaneous oil price hikes, helped throw the nation — and the world — into recession. Farmland Industries shut down some of its fertilizer plants because it could not afford to use expensive natural gas to make cheap fertilizer; many consumers were dismayed to find that their utility bills had doubled. A frenzy of new drilling resulted, which, together with a scaling back of demand due to the recession, enabled the natural gas market to recover so that prices eased back. Yet by the spring of 2001, wellhead gas prices were still twice what they had been twelve months earlier, and gas in storage had reached its lowest level ever. The nation narrowly averted serious shortages again in 2003; however, unusually mild winter and summer weather in 2004 enabled the refilling of underground gas storage reservoirs. The US has managed to avoid a train wreck so far, but given declining production, the event seems inevitable, whether it occurs this year or next.

The increasing demand for gas is coming largely from an increasing demand for electricity. To meet growing electricity needs, utilities in 2000–2001 ordered 180,000 megawatts of gas-fired power plants to be installed by 2005. This strategy seemed perfectly logical to the utilities' managers since burning gas is currently the cheapest and cleanest way to convert fossil fuel into electricity. But apparently no one in the industry had bothered to inquire whether there will be enough gas available to fire all of those new generators over their useful lifetime. Many exploration geologists are doubtful. By mid-2002, plans for many of those new gas-fired plants were being cancelled or delayed.

Does natural gas extraction follow the same Hubbert curve as does oil extraction? Oil wells are depleted relatively slowly, whereas, as we have seen, gas wells — especially newer ones — often deplete much more quickly. The typical natural gas well production profile rises from zero, plateaus for some time, and then drops off sharply. However, in aggregate, combining all of the natural gas wells in a country or large geographical region, extraction does follow a modified Hubbert curve, with the right-hand side of the curve being somewhat steeper than that for crude.

Hence, natural gas will not solve the energy-supply problem caused by oil depletion; rather, it may actually *compound* that problem. Our society is already highly dependent on natural gas and becoming more so each year. But soon we are likely to see a fairly rapid crash in production. As my colleague Julian Darley has written in his book *High Noon for Natural Gas: The New Energy Crisis*, "The coming shortage of natural gas in the United States and Canada, compounded by the global oil peak and decline, will try the energy and economic systems of both countries to their limits. It will plunge first the United States, then Canada, into a carbon chasm, a hydrocarbon hole, from which they will be hard put to emerge unscathed."[6]

Many alternative energy advocates have described natural gas as a "transition fuel" whose increased usage can enable the nation to buy time for a switch to renewable energy sources. However, in view of the precarious status of North American gas supplies, it seems more likely that any attempt to shift to natural gas as an intermediate fuel would simply waste time and capital in the enlargement of an infrastructure that will soon be obsolete anyway — while also quickly burning up a natural resource of potential value to future generations.

Coal

Currently, the US derives about as much energy from coal as it does from natural gas. Approximately 90 percent of coal mined and burned is used to generate electricity.

Coal is the most abundant of the fossil fuels, but also the most controversial one because of environmental destruction caused by coal mining, emissions from burning coal (including carbon dioxide and acid rain-causing sulphur oxides), and its inefficiency as an energy source. Coal producers typically fight all attempts to regulate emissions or to improve efficiency, and nearly all progress in these areas has come from government research in cooperation with electric utility companies.

Demand for coal has increased over the past few decades at an average pace of about 2.4 percent per year (meaning that, at current rates of increase, total usage doubles every 30 years). The EIA estimates that recoverable reserves in the US amount to about 275 billion short tons (bst), representing roughly 25 percent of total world reserves. Production in 1998 amounted to about 1.1 bst; at that rate of usage, current reserves could theoretically last 250 years. However, the EIA also notes that "much of this may not be mined because of sulfur content, unfavorable quality, mining costs and/or transportation infrastructure."

Even given these caveats, and also taking into account the fact that rates of usage are projected to continue growing, it might seem safe to assume that there are theoretically still several decades' worth of coal reserves in the US. Moreover, these reserves are already known and mapped; expensive exploration is not needed in order to locate them.

With coal, impending shortage does not appear to be as much of a problem as with oil and natural gas; however, its inefficiency, pollution, and declining net energy yield cast a pall on prospects for the increased use of coal to replace dwindling oil. Currently, we use oil to mine coal. Most of the increased coal production during the past three decades has been from opencut (open-pit) mines that are worked by relatively few miners using giant earth-moving machines that can consume as much as 100 gallons of diesel fuel per hour. As petroleum becomes less available, the energy used to mine coal will have to come from coal or some other source.

At the same time, the most easily accessed coal beds will have become depleted: like cheap oil, cheap coal relies on reserves that lie relatively close to the surface, but these represent only a small percentage of the world's total coal resources. As those are exhausted, producers will have to return to traditional underground mining. But many underground mines have been run down and allowed to flood. Moreover, most skilled miners have lost their jobs and have been routed into other occupations. Mining is difficult, dreary work, and few miners want their children to follow in their footsteps. In areas of the Western world where underground coal mining is still practiced, the average age of miners is over 40. Thus, in order to maintain or grow coal production in the future, the industry will have to find new workers as well as develop new methods of production. As this occurs, society will be deriving less net energy from the process.

In their book *Beyond Oil*, John Gever *et. al.* describe coal's depletion profile and decreasing net energy yield as follows:

> Because the United States has used only a small fraction of its total coal supply, a Hubbert analysis is only speculative ....
>
> Besides glossing over the environmental damage resulting from heavy coal use (acid rain, particulate pollution, carbon dioxide buildup in the atmosphere), optimistic projections have been based on total coal resources and have ignored the fact that substantially less net energy may ultimately be obtained from these supplies. The quality of mined coal is falling, from an energy profit ratio of 177 in 1954 to 98 in 1977 .... These estimates include only fuel used at the mine, however, and do not include the considerable amounts of energy used to build the machines used in the mines, to move the coal away from the mines, and to process it. When these costs are included, the shape of the energy profit ratio curve changes .... [and drops] to 20 in 1977... If it continues to drop at this rate, the energy profit ratio of coal will slide to 0.5 by 2040.[7]

The authors' last statement deserves some emphasis: an energy profit ratio of 0.5 means that twice as much energy would be expended in coal production as would be yielded to do useful work. Coal has a relatively low energy density to begin with, and as miners exhaust the more favorable seams and then move on, the average heat content of a pound of coal is gradually dropping. If the study by Gever and his co-authors is correct, from a net-energy standpoint *coal may cease to serve as a useful energy source in only two or three decades.*

A recently published Hubbert analysis of coal production in the US predicts that, depending on the rate of demand, production will peak between 2032 and 2060.[8]

It is theoretically possible to use coal as the raw material from which to make synthetic liquid fuels that could directly replace petroleum. The process has already been tested and used; after all, it kept the Germans going during World War II, and an improved version is currently employed by the Sasol Company in South Africal. But the net energy yield from coal-derived liquids is extremely low and will only decline further as the net energy from coal itself dwindles. Walter Youngquist writes:

> If coal were to be used in the United States as a substantial substi-
> tute for oil by liquefying it, the cost of putting in place the physical
> plants which would be needed to supply the United States with oil
> as we use it now would be enormous. And to mine the coal which
> would have to go into these plants would involve the largest mining
> operation the world has ever seen.[9]

It may be possible to improve the efficiency of the process of releasing
coal's stored energy. The most promising proposal in this regard comes from
the Zero Emission Coal Alliance (ZECA), a program started at New Mexico's
Los Alamos National Laboratory. ZECA has designed a coal power plant that
extracts hydrogen from coal and water and then uses the hydrogen to power
a fuel cell (we will discuss hydrogen and fuel cells in more detail below). The
ZECA plants would attempt to recycle nearly all waste products and heat.
Promoters claim that ZECA plants could produce electricity with an efficiency
of 70 percent, compared to an average efficiency of about 34 percent at current
combustion-based coal power plants (though newer combustion technology
already yields greater efficiencies, in the range of 55 percent). That would
mean releasing twice the energy from the same amount of coal, as compared
to the present average. ZECA's system is not truly zero-emission (no energy
production system is), but does represent a significant potential improvement
over combustion-based technologies. However, ZECA's process for the seques-
tration of CO_2 will probably constitute a significant drain on net energy yields,
and designers say the necessary fuel-cell technology is still at least five years
away from commercial application.

Abundant coal, used to generate electricity, will enable us to keep the lights
burning for a few more years; but, taking into account its other limitations —
and especially its rapidly declining net energy yield — we cannot expect it to
do much more for us in the future than it is already doing.

Nuclear Power

In a nuclear-powered electrical generating plant, uranium fuel rods are brought
together under highly controlled conditions to create an atomic chain reaction
that produces great heat. That heat is transferred to water, changing it to
steam, which turns turbines to generate electricity.

The first commercial plant built in the US was the Shippingport, Pennsylvania,
Atomic Power Station of the Department of Energy and the Duquesne Light
Company. In a dramatic high-tech dedication ceremony, ground was broken

in 1954 by President Dwight D. Eisenhower, who also opened the plant on May 26, 1958. Nuclear power was hailed as the nation's route to permanent prosperity; in reality, however, the DoE's highly touted "Atoms for Peace" program was a direct outgrowth of the nation's nuclear weapons program and served both as a public relations exercise and as a source for fissile materials for warheads.

Many nuclear power stations were built during the 1960s and '70s; today, 103 are operational in the US. In the 1950s, promoters promised that nuclear power would be so cheap as to be essentially free; but experience proved otherwise. Today, electricity from nuclear plants is inexpensive — the industry sometimes cites costs as low as two cents per kilowatt-hour — but this is true if *only* direct costs are considered. If the immense expenditures for plant construction and safety, reactor decommissioning, and waste storage are taken into account, nuclear power is very expensive indeed.

During the 1970s and '80s, an antinuclear citizens' movement was successful in swaying public opinion against nuclear technology and in discouraging the further growth of the industry. The movement's warnings about the dangers of nuclear power were underscored by serious reactor accidents at Three Mile Island in Pennsylvania and Chernobyl in the Soviet Union; other less-publicized accidents have plagued the industry from its inception and continue to do so. As a result of both greater-than-anticipated expenses and public wariness, no orders for new plants have been placed in the US since the 1970s.

Nuclear power plants produced 3.6 percent of all the energy consumed in the US in 1980; by 2000, that number had climbed to 8.1 percent. This increase was due not to the building of new reactors, but to increased efficiency in the operation of existing plants. In 2000, the industry achieved a record overall average capacity factor (the percentage of potential output actually achieved on average) of nearly 86 percent, up from 58 percent 20 years earlier.

Today about 20 percent of all the electricity generated in the US comes from nuclear sources. Globally, 12 percent of the world's electricity, and 5 percent of the total energy consumed, are nuclear-generated. Some nations derive much more of their energy from nuclear plants than does the US: France, for example, gets 77 percent of its electricity from atomic energy, Belgium 56 percent, and Sweden 49 percent. There are currently 442 reactors operating worldwide. In Western Europe, France is the only country still building nuclear plants; only in Asia is the nuclear-power industry expected to expand significantly in the foreseeable future.

Could nuclear power take up the slack as energy from petroleum production declines? Those who argue that it could claim that nuclear power is:

Abundant: There is a virtually limitless supply of fuel (assuming breeder reactors, which reprocess spent fuel);

Clean: It is non-polluting, having no CO_2 emissions; wastes are produced in small quantities and the problem of their disposal will be solved once a single permanent repository is created;

Practical: Nuclear fuel has the highest energy density of any fuel known; further, nuclear power is inexpensive, the produced electricity being cheaper than energy from coal; and

Safe: It is safer than many people believe, and becoming safer all the time. The likelihood of a person dying from a nuclear accident is already far lower than that of dying in an airplane crash, while new technology on the drawing boards will make nuclear power virtually 100 percent safe in the future.

However, when these claims are examined in detail, a very different picture emerges.

Abundant? The fuel supply for nuclear power is virtually limitless if we use fast-breeder reactors to produce plutonium — which is one of the most poisonous materials known and is used to make nuclear weapons. But only a few fast-breeder reactors have been constructed, and they have proved to be prohibitively expensive, largely as a result of the need for special safety systems.

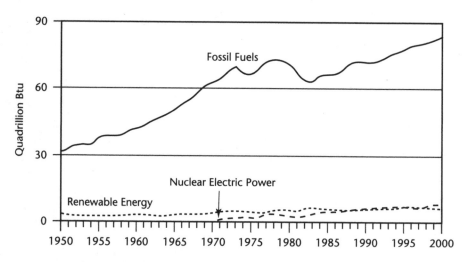

Figure 19. US energy consumption by source

(Source: US Energy Information Administration)

These reactors generate an extraordinary amount of heat in a very small space and use molten metals or liquid sodium to remove the heat. Designing reactors to take these properties into account has made them costly to build and maintain. It also makes them susceptible to serious fires and long shutdowns: the French Superphoenix reactor operated for less than one year during the first ten years after it had been commissioned.

France and the UK, despite having pursued breeder programs for several decades, have no plans for constructing more such plants. Japan has not restarted its Monju reactor, which was shut down after a sodium fire in December 1995. Among countries that have constructed breeders, Russia alone supports further development.

It is also possible to reprocess spent fuel into a form known as MOX (mixed oxide), which consists of a mixture of plutonium and uranium oxides. Reprocessed MOX fuel can then be used to replace conventional uranium fuel in power plants. However, only two MOX plants have been built (one in the UK, the other in France), and both have turned out to be environmental and financial nightmares.[10]

Uranium — the usual fuel for conventional reactors — must be mined, and it exists in finite quantities. The US currently possesses enough uranium to fuel existing nuclear reactors for the next 40 years.[11] The mining process is wasteful, polluting, and dangerous: the early New Mexico uranium mines, which employed mostly Navajo workers, ruined thousands of acres of Native lands and poisoned workers and their families. The entire episode constitutes a horrific and permanent blot on the industry's record.[12]

Further, much of the energy needed to mine uranium currently comes from oil. As petroleum becomes more scarce and expensive, the mining process will likewise become more costly and will yield less net energy.

Clean? Vice President Dick Cheney told CNN on May 8, 2001, that nuclear power "doesn't emit any carbon dioxide at all."[13] But this is true only in the sense that the nuclear chain reaction itself doesn't create such emissions. Mining uranium ore, refining it, and concentrating it to make it fissionable are all highly polluting processes. If the whole fuel cycle is taken into account, nuclear power produces several times as much CO_2 as renewable energy sources.

The assertion that nuclear waste is only produced in small quantities is misleading. Direct wastes include roughly 1,000 metric tons of high- and low-level waste per plant per year — hardly a trivial amount, given that much of this waste will pose hazards for thousands or tens of thousands of years to come. Further-more, this figure does not include uranium mill tailings, which are also radioactive and can amount to 100,000 metric tons per nuclear power plant per year.[14]

Can the problem of nuclear waste be solved by the creation of a permanent repository? To assume so is to indulge in wishful thinking. After nearly five decades of the development and use of atomic energy, no country in the world has yet succeeded in building a permanent high-level nuclear waste repository. Moreover, the transporting of wastes to such a central repository would create extra dangers.[15]

Practical? It is true that nuclear fuel has an extraordinarily high energy density, but this is the case only for uranium that has already been separated from tailings and been processed — which itself is a far more hazardous and energy-intensive procedure than drilling for oil or mining coal.

The costs typically quoted for nuclear-generated electricity (1.8¢-2.2¢/ kWh) are operating costs only, including fuel, maintenance, and personnel. As noted earlier, such figures omit costs for research and development, plant amortization and decommissioning, and spent-fuel storage. Fully costed, nuclear power is by far our most expensive conventional energy source. Indeed, total costs are so high that, following the passage of energy deregulation bills in several states, nuclear plants were deemed unable to compete, and so utility companies like California's PG&E had to be bailed out by consumers for nuclear-related "stranded costs."[16] Germany has decided to phase out nuclear power for both economic and environmental reasons.

If nuclear energy is not cheap, is it at least reliable? Certainly more so than it was two or three decades ago. However, it is worth noting that problems at the Diablo Canyon and San Onofre reactors contributed significantly to California's energy crisis in 2001. Nuclear power plants are extremely complex — many things can go wrong. When technical failures occur, repair costs can be much higher than is the case with other types of generating plants.

Safe? For the general public, safety is probably the foremost concern about nuclear power. Siting nuclear plants has always been a challenge, as communities typically fear becoming the next Three Mile Island or Chernobyl. Earthquake zones must be ruled out, along with most urban areas (due to evacuation problems). While the statistical likelihood of any given individual dying in a nuclear accident is quite low, if a truly catastrophic accident were to occur many thousands or even millions could be sickened or die as a result. Nuclear power's record of mishaps is long and disturbing. It is a telling fact that the industry has required special legislation (the Price-Anderson Act) to limit the liability of nuclear-power plant operators in the event of a major accident. If the technology were as safe as that in conventional generating plants, no such measure would be needed. Following the terrorist attacks of September 11, many commentators pointed out that if the airplane hijackers had targeted nuclear power plants rather than office buildings, the resulting human toll would have been vastly greater.

Extraordinary safety claims have been made for a new design of high-temperature reactor, the Pebble Bed Modular Reactor. However, this technology is strictly theoretical, never yet having been tested in practice. Even the International Atomic Energy Agency's International Nuclear Safety Advisory Group has expressed misgivings about claims that the ceramic coating of the fuel "pebbles" can take the place of a normal reactor containment building. This coating consists mostly of graphite; and though graphite has a very high melting point, it can burn in air (graphite burned in the Chernobyl disaster as well as in the 1957 Windscale fire), so it is important to exclude air from the reactor. Current assertions that these untested technologies will be "100 percent safe" are probably about as believable as claims made in the 1950s that nuclear-generated electricity would be "too cheap to meter."[17]

These are all important concerns in assessing to what extent the deployment of nuclear power has been successful or even acceptable so far. But in deciding whether this energy source can help us through the transition away from oil and natural gas, we need to consider three other questions: Can the technology be scaled up quickly enough? What is its EROEI? And to what extent can it substitute for petroleum in the latter's current primary uses, such as in transportation and agriculture?

Scaling up the production of electricity from nuclear power would be slow and costly. In the US, just to replace current electricity generated by oil and natural gas, we would need to increase nuclear power generation by 50 percent, requiring roughly 50 new plants of current average capacity. But this would do nothing to replace losses of energy to transportation and agriculture as petroleum becomes less available.

Since coal is currently used mostly for electricity generation, nuclear power could conceivably substitute for coal; in that case, nuclear generation would have to increase by 250 percent — requiring the construction of roughly 250 new atomic power plants.

But using atomic energy as a replacement for petroleum is much more problematic. To replace the total amount of energy used in transportation with nuclear-generated electricity would require a vast increase (on the order of 500 percent) in nuclear generation capacity. Moreover, the replacement of oil — gasoline, diesel, and kerosene — with electricity in the more than 700 million vehicles worldwide constitutes a technical and economic problem of mammoth proportions. Current storage batteries are expensive, they are almost useless in very cold weather, and they need to be replaced after a few years of use. Currently, there are no batteries available that can effectively move heavy farm machinery or propel passenger-carrying aircraft across the oceans. (We will return to the problem of storing electrical energy later in this chapter, in discussions about hydrogen and fuel cells.)

Finally, the EROEI for nuclear power — when plant construction and decommissioning, waste storage, uranium mining, and all other aspects of production are taken into account — is fairly low. Industrial societies have, in energy terms, been able to afford to invent and use nuclear technologies primarily because of the availability of cheap fossil fuels with which to subsidize the effort.

For all of these reasons, it would be a disastrous error to assume that nuclear power can enable us to maintain business as usual when energy shortages arise due to the depletion of fossil fuels. New nuclear plants will no doubt be proposed and built as energy shortages arise; however, the associated costs will be too high to permit the construction of enough plants, and quickly enough, to offset the decline of cheap fossil fuels.

Wind

As we saw in Chapters 1 and 2, the capture of energy from wind — first by sails for transportation over water, and then by mills used to grind grain or pump water — predates industrialism. Today, sleek high-tech turbines with airplane propeller-like blades turn in response to variable breezes, generating an increasing portion of the world's electricity.

Winds arise from the uneven heating of the Earth's atmosphere by the Sun, as well as from Earth's surface irregularities and its axial rotation. Winds are generally strongest in mountain passes and along coastlines. The world's best coastal wind resources are in Denmark, the Netherlands, California, India, southern Argentina, and China; "wind farms" have been developed in all of these places.

Wind is a limited but renewable energy resource: unlike fossil fuels, winds are not permanently "drawn down" by their use. Once a wind turbine is installed, costs are incurred primarily for its maintenance; wind itself is, of course, free.

Of all renewables, wind is the one that, on a global level, is being developed the fastest. Wind power is approaching 40 gigawatts in installed capacity worldwide, out of the total electrical generating capacity of 3000 gW. Germany and Spain have recently become the world leaders in installed wind generating capacity. In the US, growth in the industry slowed in the 1990s but began a resurgence in 2000; about one percent of all electricity generated in the nation now comes from wind.

Wind-turbine technology has advanced dramatically in the past few years. Only a decade ago, engineers envisioned turbines with a maximum capacity of 300 kW, and blade rotation speeds were such that many areas had to be excluded from siting consideration for environmental reasons (turbine blades sometimes kill endangered birds, which tend to migrate along coastal areas). The optimum wind speeds for the turbines produced then were 15 to 25 MPH and only about 20 percent of actual wind energy could be converted to electricity.

Turbines that are being developed and installed today have capacities in the range of two to three megawatts. Blade rotation is much slower (resulting in less likelihood of bird kill), and efficiencies have been improved significantly. Moreover, the newer turbines can operate in more variable winds — with speeds ranging from about 7 to 50 MPH.

The cost of wind-generated electrical power is declining quickly. The National Renewable Energy Laboratory (NREL) estimates that by 2010 average prices will be in the range of 3.5¢/kilowatt-hour. The Lake Benton Wind Farm in

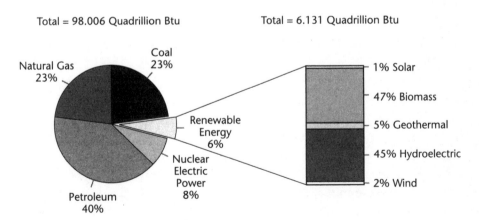

Figure 20. US energy consumption by source, showing renewables, 2003

(Source: US Energy Information Administration)

Minnesota, operational as of 2002 and using 1 mW turbines, produces wind-generated electrical power at 3.2¢/kWh. Another large project, on the Oregon/Washington border, is expected to produce power at 2.5¢/kWh. These prices are already competitive with other generating sources; and as the EROEI of coal declines and natural gas supplies dry up, wind power will look even more inviting.

New vertical-axis turbine designs being developed at Lawrence Berkeley Laboratories in cooperation with the Makeyev State Rocket Center in Miass, Russia, could make wind power more feasible in a wider range of situations. Prototypes feature vertical fiberglass blades that rotate around a central mast. The company that has been formed to commercialize the design, Wind Sail, expects to market small turbines to homeowners. Previous horizontal-axis designs were noisier and had a tendency to kill birds — problems solved by the new design. Vertical-axis turbines are also potentially more efficient than similar-sized horizontal-axis turbines.[18]

How much energy could be derived from wind? Theoretically, a great deal. A good guide is a 1993 study by NREL that concluded that about 15 quads (quadrillion BTU) of energy could be produced in the US per year. Since the newer turbines are capable of operating in a wider range of wind conditions, that potential could conceivably now be in the range of 60 quads. Total energy usage in the US is about 100 quads.[19]

However, the realization of that potential will require huge investments and a strong commitment on the part of policymakers. Investment will be required not just for the turbines themselves, but also for new transmission lines: a 1991 California study estimated that only 12 percent of the "gross technical potential" for wind power in that state could be realized given the existing transmission infrastructure.

In addition, it will be necessary to solve technical problems arising from wind power's intermittent daily, monthly, and seasonal availability. Often, peak availability of wind does not correspond with peak energy demand. This is not an insurmountable problem: energy storage systems (such as the Regenesys regenerative electrochemical fuel cell) are in development that may in the future eliminate the daily variability of electricity generation from wind.[20] Also, peak wind generation that exceeds momentary demand could be used to produce hydrogen (see 167).

Over the short term, the problem of intermittency should not simply be shrugged off. Germany, which now leads the world in installed wind electrical generation capacity (14,350 Megawatts at the end of 2003), therefore also has the most experience with the practical problems associated with wind energy. A recent report from EON, the largest grid operator in Germany, points out that it is necessary to have 80 percent of wind capacity available at all times from power stations that can produce on-demand energy (i.e., coal, nuclear, hydro, geothermal, or natural gas plants). In addition, according to the report, "if wind power forecast differs from the actual infeed, the transmission system operator must cover the difference by utilizing reserve capacity. This requires reserve capacities amounting to 50 to 60 percent of the installed wind capacity." The report's authors also point out that wind power often requires the construction of new grid capacity to transport the electricity from remote areas, where the wind farms operate, to populated areas where the electricity is consumed.[21]

Though the siting of wind turbines presents a challenge, imaginative solutions are being proposed. Most of the best sites are privately owned and in use for other purposes — principally, for agriculture. However, wind turbines do not take up exorbitant amounts of space, and wind farms and conventional farms need not be mutually exclusive. A Minnesota farmer earning less than $30 per acre per year from livestock and $250 per acre from crops might earn $1,000 per acre from land rental for a wind farm and continue to use most of the land for cattle or corn.

At the moment, the EROEI for wind is the best for any of the renewables that has much opportunity for expansion. While Odum gives a figure of 2+, a Danish study suggests an energy payback period of only two to three months, which might translate to an EROEI of 50 or more.[22] Though even the latter number may be relatively low when compared to the EROEI for oil and natural gas during the expansion phase of industrial civilization (when it occasionally surpassed 100-to-1), it probably already exceeds the EROEI for these fossil fuels as their net energy yield gradually wanes due to depletion.

Wind can deliver net energy; the challenge for industrial societies is to scale up production quickly enough to make up for the energy decline from dwindling oil and natural gas supplies. Just to produce 18 quads of wind power in the US by 2030 (never mind the 60 quads of theoretical potential) would require the installation of something like half a million state-of-the-art turbines, or roughly 20,000 per year starting now. That is five times the present world production capacity for turbines. This feat could be accomplished, but it would require a significant reallocation of economic resources. Meanwhile, most of the energy needed for that undertaking would have to come from dwindling fossil fuels.

Thus even if current policymakers had the political will to undertake such a transition, industrial societies would still face a wrenching adjustment to a lower-energy regime. This sobering assessment is underscored by the difficulty of substituting wind-generated electricity for oil's current uses. As we saw in the previous section on nuclear power, electricity is not well suited to the powering of our current transportation and agriculture infrastructure. The rebuilding of that infrastructure is itself a gargantuan task in both economic and energy terms, and one that is still beset by technical challenges.

Nevertheless, it is clear that, of the alternatives we have surveyed so far, wind is probably the most practicable.

Solar Power

Since virtually all terrestrial energy sources derive ultimately from the Sun, the development of direct means of capturing usable energy from sunlight seems an obvious way to satisfy industrial societies' prodigious appetites for power. There is, after all, plenty of solar energy available: the average solar energy influx in North America is about 22 watts per square foot (200 watts per square meter), which means that the typical suburban house in the US continuously receives the equivalent of over 25 horsepower in energy from the Sun. However, there are technical obstacles to gathering that energy, converting it to useful forms, and storing it for times when the Sun is not shining.

Solar energy is most easily harvested and used in the form of heat. For millennia, people have oriented their homes to take advantage of the Sun's warming rays; today, the design of houses to maximize passive solar heating is still one of the most effective ways to increase energy efficiency. Simple rooftop collectors for home hot water or swimming pool heating also take advantage of free solar heat.

The ancient Greeks and Chinese used glass and mirrors to focus the Sun's rays in order to start fires. Modern solar-thermal electrical generation technologies use the same principle to produce electrical power by heating water or other fluids to temperatures high enough to turn an electrical generator. Several distinct types of solar-thermal generating systems have been developed (including dish concentrators driving Stirling engine generators; trough concentrators heating a liquid-to-gas system driving a turbine generator; solar towers using large reflector arrays to heat molten salts which, through a heat exchanger, drive steam turbines; and plastic film collectors that work much like trough concentrators, but are much cheaper to build). Relatively few such systems of any type are in use, but ambitious plans are on the drawing boards, including some that integrate solar-thermal systems into the roofs of commercial and industrial buildings.

The photovoltaic effect, in which an electrical current is directly generated by sunlight falling upon the boundary between certain dissimilar substances, was discovered in 1839 by a nineteen-year-old French experimental physicist named Edmund Becquerel. Albert Einstein won the Nobel Prize in 1923 for explaining the effect. The first silicon solar-electric cells were made in the 1950s by researchers at Bell Laboratories, who achieved an initial conversion efficiency of only 4.5 percent. The development of photovoltaic (PV) technologies soon received a significant boost from research undertaken by the US space program, which used solar cells to power satellites. By 1960, efficiencies had been boosted to nearly 15 percent. In the 1970s, alternative energy enthusiasts began to envision a solar future in which photovoltaics would play a significant role in powering a post-petroleum energy regime.

Today there is roughly 1 gW of PV generating capacity installed worldwide (versus roughly 3000 gW of capacity in conventional power plants). Power-conversion efficiencies are now as high as 30 percent, and the cost of solar cells — initially astronomical — has fallen a hundred-fold. A typical small system now costs as little as $6 per watt of production capacity, whereas on large-scale projects costs as low as $3 are possible; at the latter price, with financing of the system at 5 percent interest over 30 years, the price of produced PV electricity amounts to roughly 11¢/kWh — though few installations actually achieve such a low cost. Photovoltaic electricity is still expensive.

PV technologies have the advantage of being able to provide electricity wherever there is sufficient sunlight, so they are ideal for powering remote homes or villages that are difficult to connect to a power grid. With a PV system, homeowners can become independent of electrical utility companies altogether. The disadvantage of such "stand-alone" systems is that a means must be provided to store electrical power for use when the Sun isn't shining — at night or on cloudy days. The typical solution is a bank of batteries, which require maintenance and add substantially to the system's cost. A complete system normally includes a collector array, a controller, an inverter (to change the generated current from DC to AC), and a battery bank, which altogether may represent an investment of more than $20,000 for even an energy-conserving home. In many states, businesses and homeowners can tie their PV panels directly to a power grid; by doing so, they avoid both electric bills and the need for batteries (though an inverter is still required). In this case, the system owner becomes an independent commercial electricity generator, selling power to the local utility company. Such grid-tied systems are typically much less expensive than stand-alone systems.

Two technical improvements in PV technology that are now in the developmental stage — thin-film panels and PV dye coatings — seem especially promising for reducing the cost of photovoltaic electricity. To date, the biggest obstacle to further implementation of the technology has been that production costs are high. The fabrication of even the simplest semiconductor cell is a complex process that has to take place under exactly controlled conditions, such as a high vacuum and temperatures between 750 and 2550 degrees Fahrenheit (400 and 1400 degrees Celsius). These new technical improvements promise to lower production costs dramatically.

Researchers are now experimenting with the use of hybrid materials that are inexpensive and allow for the use of flexible substrates, such as plastics. Manufacturers of such thin-film PV collectors claim a possible production cost of electricity of 7¢/kWh. There are three forms of thin-film PV technology in commercial production: amorphous silicon (a-Si), cadmium telluride (CdTe), and copper indium diselenide (CuInSe2, or CIS). There are two more on the

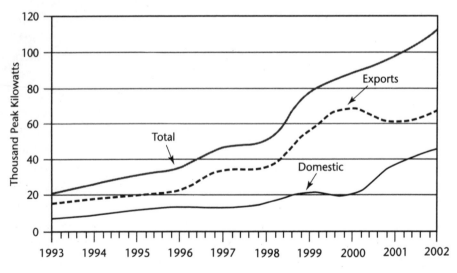

Figure 21. Shipments of PV cells and modules, 1993 – 2002

(Source: International Energy Agency)

way: spheral and CIGS (copper indium gallium diselenide).[23] Already, amorphous silicon accounts for more than 15 percent of the worldwide PV production. Amorphous silicon technology holds great promise in building-integrated systems, replacing tinted glass with semi-transparent modules; however, the efficiency is low: while some experimental a-Si modules have exceeded 10 percent efficiency, commercial modules operate in the 5 to 7 percent range. Cadmium Telluride laboratory devices have approached 16 percent efficiency, though production modules have achieved only about 7 percent. Copper Indium has reached a research efficiency of 17.7 percent, with a prototype power module reaching 10.2 percent, but production problems have so far prevented any commercial development.

Meanwhile, scientists at Switzerland's École Polytechnique de Lausanne have developed a fundamentally different solar pholtovoltaic cell that may eventually result in the cheapest PV devices of all. The production process uses common materials and low temperatures: a photosensitive dye, whose properties enable it to perform what the technology's promoters call "artificial photosynthesis," is simply silkscreened onto a substrate, such as glass. The resulting cells, known as Titania Dye Sensitised Cells (Titania DSC), can be assembled into colored opaque or translucent modules that could potentially be incorporated into the walls of buildings or the sunroofs of cars. Titania DS cells demonstrate performance in low light and at high temperatures that far surpasses that of silicon cells. Titania cells are currently only 10 percent efficient in energy conversion.[24] In this case, lower efficiency (relative to silicon-crystal cells) may not be much of a problem because of the potential for enormous cost savings: it may not matter much if a solar cell is inefficient if it can be put where otherwise only tarpaper, a sheet of plywood, or glass would go.

Nanosolar, a startup company in Palo Alto, California, is planning to commercialize this new technology, with its first product slated to hit the market in 2006. The production process will involve spraying a combination of alcohol, surfactants (substances like those used in detergents), and titanium compounds on a metal foil. A *Technology Review* article describes what happens next: "As the alcohol evaporates, the surfactant molecules bunch together into elongated tubes, erecting a molecular scaffold around which the titanium compounds gather and fuse. In just 30 seconds, a block of titanium oxide bored through with holes just a few nanometers wide rises from the foil. Fill the holes with a conductive polymer, add electrodes, cover the whole block with a transparent plastic, and you have a highly efficient solar cell."[25] Nanosolar hopes to reduce the cost of solar electricity by up to two thirds, making it competitive with commercial grid electricity rates. Eventually, it may be possible to paint a photovoltaic material directly onto buildings, cars, and other objects.

Still another new solar photovoltaic technology, this one involving organic materials, was recently announced by researchers at the Georgia Institute of Technology.[26] Using a crystalline organic film, pentacene, together with C_6O, a form of carbon more popularly known as "buckyballs," the research group was able to convert sunlight into electricity with 2.7 percent efficiency, and they hope to reach 5 percent efficiency in the near future. Though the efficiency of the material is likely to remain low, its flexibility and minimal weight would allow it to be used on nearly any surface, including tents and clothing. The developers estimate that commercial residential applications are five years away, though versions to power small devices could be marketed within two years.

Net-energy calculations for current photovoltaic technologies are a matter of some controversy. Clearly, conventional silicon-crystal cells have so far had a relatively low return for the energy invested in their manufacture, even though promoters of the technology staunchly claim a favorable figure (typically, they exclude from their analyses the energy expended in transportation as well as that embodied in production facilities). In this instance at least, net-energy payback appears to be highly sensitive to the volume of production: PV modules are still manufactured on a very small scale; if demand were to surge, the energy returned on investment would likely rise very noticeably. It is likely that, even if the most pessimistic assessments of silicon-crystal cells — which suggest a *current* net return of less than 1:1 — are correct, the newer thin-film and DSC technologies may be able to achieve a substantially more favorable EROEI (the more optimistic assessments of silicon-crystal cells suggest a current net return of roughly 10).[27] At some point the net energy available from PV electricity will overtake the EROEI that can be derived from petroleum, as the latter is depleted.

However, solar photovoltaic and thermal-electric technologies present us once again with the problem we noted concerning nuclear power and wind: electricity cannot easily be made to power our current transportation and agriculture infrastructure. What is needed is some efficient medium for storing electrical energy that also renders that energy transportable and capable of efficiently moving large vehicles.

Many people believe that the solution lies in the simplest and most abundant element in the universe.

Hydrogen

Hydrogen is the lightest element, and it combines readily with oxygen; when it does so, it burns hot; and its combustion product is water — no greenhouse gases, no particulate matter or other pollutants. For these and other reasons, hydrogen would seem to be an attractive alternative to fossil fuels.

However, there are no exploitable underground reservoirs of hydrogen. Usable hydrogen has to be manufactured from hydrocarbon sources, such as natural gas or coal (a gallon of gasoline actually contains more hydrogen than does a gallon of liquid hydrogen), or extracted from water through electrolysis. Hydrogen production from algae and from sewage wastes has been demonstrated in the laboratory, but it is unclear whether these processes can ever be scaled up for commercial application. The crux, however, is this: *The process of hydrogen production always uses more energy than the resulting hydrogen will yield*. Hydrogen is thus not an energy *source*, but an energy *carrier*.

Still, many people foresee a prominent role for hydrogen as a means to enable renewable wind- and photovoltaic-generated electricity to be stored and transported. Proposals for a "hydrogen economy" have been circulating for decades (a 1976 study by the Stanford Research Institute was entitled *The Hydrogen Economy: A Preliminary Technology Assessment*), and in recent years a chorus of proponents has proclaimed the desirability and inevitability of a full transition from fossil fuels to an energy regime based on renewables and hydrogen. "Hydrogen-powered fuel cells promise to solve just about every energy problem on the horizon," writes David Stipp in an article called "The Coming Hydrogen Economy."[28] At the Hyforum held in Munich, Germany, in September 2000, T. Nejat Vezirogllu, President of the International Association for Hydrogen Energy, proclaimed, "It is expected that the petroleum and natural gas production fueling this economic boom will peak around the years 2010 to 2020 and then start to decline. Hydrogen is the logical next stage, because it is renewable, clean, and very efficient."[29]

Much of the optimism surrounding the hydrogen-economy vision — whose boosters occasionally exhibit a techno-utopianism of almost messianic intensity — derives from recent developmental work on fuel cells, which chemically produce electrical energy from hydrogen without burning it. Fuel cells have more in common with batteries than with combustion engines.

Hydrogen is not the only substance that can be used to power fuel cells. The Regenesys fuel cell uses two electrolyte salt solutions; it will be useful alongside conventional and renewable commercial power plants to store output and

Nonrenewable and Renewable Energy Sources

Nonrenewable	Renewable
Oil	Hydroelectric
Natural Gas	Wind
Coal	Solar Power
Nuclear Power	Biomass, including biodiesel and ethanol
Geothermal Power (geysers)	Tides
	Waves
	Geothermal (ground-water heat pumps)

Net Energy Compared

Below are the summarized results of two comprehensive comparative studies of net energy (EROEI), one by Cleveland, Costanza, Hall, and Kaufmann (1984), the other by Odum (1996). Cleveland and Kaufmann have criticized Odum's methodology (see www.oilanalytics.com), but have not published an updated study of their own. Time is relevant to EROEI studies because the net-energy yield for a given energy source may change with the introduction of technological refinements or the depletion of a resource base.

Process	Energy Profit Ratio
Nonrenewable	
Oil and gas (domestic wellhead)	
1940s	Discoveries > 100.0
1970s	Production 23.0, discoveries 8.0
Coal (mine mouth)	
1950s	80.0
1970s	30.0
Oil shale	0.7 to 13.3
Coal liquefaction	0.5 to 8.2
Geopressured gas	1.0 to 5.0
Renewable	
Ethanol (sugarcane)	0.8 to 1.7
Ethanol (corn)	1.3
Ethanol (corn residues)	0.7 to 1.8 ☞

Process	Energy Profit Ratio
Methanol (wood)	2.6
Solar space heat (fossil backup)	
Flat-plate collector	1.9
Concentrating collector	1.6
Electricity production	
Coal	
US average	9.0
Western surface coal	
No scrubbers	6.0
Scrubbers	2.5
Hydropower	11.2
Nuclear (light-water reactor)	4.0
Solar	
Power satellite	2.0
Power tower	4.2
Photovoltaics	1.7 to 10.0
Geothermal	
Liquid dominated	4.0
Hot dry rock	1.9 to 13.0

(Source: From C. J. Cleveland, R. Costanza, C. A. S. Hall, and R. Kaufmann, "Energy and the U.S. Economy: A Biophysical Perspective," *Science* 225 (1984), pp. 890-97.)

Item	Energy Yield Ratio
Dependent Sources, No Emergy Yield	
Farm windmill, 17 MPH wind	0.03
Solar water heater	0.18
Solar voltaic cell electricity	0.41
Fuels, Yielding Net Emergy	
Palm oil	1.06
Energy-intensive corn	1.10
Sugarcane alcohol	1.14
Plantation wood	2.1

Item	Energy Yield Ratio
Lignite at mine	6.8
Natural gas, offshore	6.8
Oil Mideast purchase	8.4
Natural gas, onshore	10.3
Coal, Wyoming	10.5
Oil, Alaska	11.1
Rainforest wood, 100 years growth	12.0
Sources of Electric Power, Yielding Net Emergy	
Ocean-thermal power plant	1.5
Wind electro-power	2-?
Coal-fired power plant	2.5
Rainforest wood power plant	3.6
Nuclear electricity	4.5
Hydroelectricity	10.0
Geothermal	13.0
Tidal electric, 25 ft. tidal range	15.0

(Source: From Howard T. Odum, *Environmental Accounting, Emergy, and Decision Making* (John Wiley, 1996). Note: In that book, Odum explains the meaning of his term "emergy." If you think of it as shorthand for "embodied energy," you will not be far from the mark, though Odum's technical definition of the term is far more rigorous and complicated.) ∎

release it when needed. In addition, zinc-air fuel cells are in development which, if the promotional literature is to be believed, are much cheaper to make than hydrogen fuel cells, use a solid fuel that has twice the energy density of hydrogen, and have an electricity-to-electricity efficiency in the range of 40 to 60 percent.[30] Zinc "fuel" will come in the form of small pellets. The chemical reaction in zinc fuel cells produces zinc oxide, a non-toxic white powder. When all or part of the zinc has been transformed into zinc oxide, the user refuels the cell by removing the zinc oxide and adding fresh zinc pellets and electrolyte. The zinc oxide is then reprocessed into new zinc pellets and oxygen in a separate, stand-alone recycling unit, using electrolysis. Thus, the process is a closed cycle that can theoretically be continued indefinitely. Each cycle consumes energy; but we must remember that the real purpose of the fuel cell is not to *produce* net energy, but rather to make stored energy available for convenient use.

But back to hydrogen. At present, on a global scale, about 40 million tons of hydrogen are produced commercially per year. This represents slightly more than one percent of the world's energy budget. Most of this commercially produced hydrogen is now made from natural gas.

There are reasons to be hopeful about hydrogen's potential. The electric drive train of a fuel cell-driven car would be much lighter than a conventional gasoline or diesel drive train. Emissions from burning hydrogen in fuel cells consist only of water and heat; thus many pollution problems — including the production of greenhouse gases — could be reduced dramatically by the widespread use of hydrogen. Even if the source of hydrogen is coal or natural gas, fewer emissions are produced in the coal or gas reformation process (the production of hydrogen) than in the direct burning of these fossil fuels for energy.

Several major car manufacturing companies are currently working on new models that will run on hydrogen fuel cells. The experimental Daimler-Benz NECAR 3 (New Electric Car, version 3), for example, generates hydrogen on-board from methanol — thus dispensing with the problematic extra weight of batteries and hydrogen tanks. Another solution to the weight problem is to redesign the entire automobile for maximum weight reduction and aerodynamics; this is the approach taken by the "Hypercar," a project of Hypercar Inc.[31]

Hydrogen production is also being proposed as a means to store electrical energy from solar panels or wind turbines in homes or commercial buildings, replacing bulky and inefficient batteries. Hydrogen-powered fuel cells could thus enable a transition to decentralized energy production, reducing costs for the construction and maintenance of centralized generating plants and transmission lines.

Amory Lovins of the Rocky Mountain Institute has published "A Strategy for the Hydrogen Transition," illustrating how "the careful coordination of fuel-cell commercialization in stationary and transportation applications, the use of small-scale, distributed fueling appliances, and Hypercars combine to offer leapfrog opportunities for climate protection and the transition to hydrogen."[32] Implicit in the plan is a reliance on natural gas as the primary source for hydrogen for at least two decades, until renewable energy souces can be scaled up.

That's the good news about hydrogen. Unfortunately, there is bad news as well.

A hydrogen energy infrastructure would be quite different from our present energy infrastructure, and so the transition would require time and the investment of large amounts of money and energy. That transition would be aided tremendously if we were to switch present government subsidies from nuclear power, oil, and coal to renewables, fuel cells, and hydrogen. But, given the political influence of car and oil companies and the general corruption and inertia of the political process, the likelihood of such a subsidy transfer is slim for the moment. Yet if we simply wait for price signals from the market to trigger the transition, it will come far too late.

An even greater problem is the current and continuing reliance on natural gas for hydrogen production. Hydrogen proponents assume the continued, abundant availability of natural gas as a "transition fuel." Without some transitional hydrocarbon source, there is simply no way to get to a hydrogen economy: there is not enough net energy available from renewable sources to "bootstrap" the process while supporting other essential economic activity. As we have seen, prospects for maintaining — much less increasing — the natural gas supply in North America appear disturbingly uncertain. Within only a few years, decision makers will be confronting the problem of prioritizing dwindling natural gas supplies — should they fund the transition to a hydrogen economy or heat people's homes during the winter? Faced with a crisis, they would find it difficult to justify diverting natural gas supplies away from immediate survival needs.

In terms of energy efficiency (setting aside for the moment the problem of emissions and the need for energy storage), we would be better off burning natural gas or using PV or wind electricity directly, rather than going through the extra step of making hydrogen. The Second Law of Thermodynamics insures that hydrogen will be a net-energy loser every time since some usable energy is lost whenever it is transformed (e.g., from sunlight to photovoltaic electricity, from electricity to hydrogen, or from hydrogen back to electricity).

Given the already low net energy from renewables as well as the net energy losses from both the conversion of electricity to hydrogen and the subsequent conversion of hydrogen back to electricity, it is difficult to avoid the conclusion that the "hydrogen economy" touted by well-meaning visionaries will by necessity be a much lower-energy economy than we are accustomed to.

The future may well hold hydrogen fuel cell-powered cars — but not in numbers approaching the current global fleet of 775 million vehicles. In the low-energy social environment toward which we are inevitably headed, it will be possible for only a tiny wealthy minority to navigate over disintegrating streets and highways in sophisticated, highly efficient Hypercars. For the rest of us, a good pair of shoes and a sturdy bicycle will be the best affordable transport tools.

I recently toured the Schatz Energy Research Center (SERC) at Arcata, California, one of the nation's foremost research centers for hydrogen, fuel cells, and renewable energy. The mission of the center is to promote the use of clean and renewable energy. The Schatz lab, housed in a small, converted 1920s hospital building, specializes in generating hydrogen fuel from solar photovoltaics. The lab designed and built a 9kW fuel cell powered car based on a small European electric vehicle — the first street-ready fuel-cell car in the US. SERC has also made a fuel cell that powers a microwave relay station providing telephone service for the Yurok Tribe of Northern California.

Peter Lehman, the SERC Director, showed me several bench-top, state-of-the-art fuel cells — each handmade and expensive to build. Lehman said that for most small-scale applications (including homes and personal automobiles), batteries are still a more efficient storage medium for energy than hydrogen. In most cases, according to Lehman, it just doesn't make sense to take high-quality energy in the form of electricity, turn it into hydrogen, and then turn it back into electricity, since there are losses at each stage along the way — if there are ways of using the electricity directly. However, in larger-scale generation situations — say, a wind farm — at times when there is no immediate use for the electricity being generated, hydrogen production could provide a way to store energy while also producing a transportation fuel for fuel cell vehicles such as trucks or buses. But in commuting situations, when mileage requirements are low, Lehman feels that battery electric vehicles are more efficient and the right choice for private cars. In the foreseeable future, gasoline or diesel hybrid cars also make more sense than do fuel cell vehicles.

The two biggest problems with fuel cells currently, according to Lehman, are that they don't last long enough, and they're expensive. Schatz's cells are now able to perform for about 2,000 hours (that's three months of continuous operation). The upside is that fuel cells can be remanufactured, so that a user could rotate two cells, with one on the job while the other is being refurbished. But this would, of course, increase the already daunting cost. The Schatz lab is working to overcome both these limitations, but Lehman admits that there is a long way to go, and advances appear to be incremental and slow. There is currently no off-the-shelf, production-model fuel cell available anywhere that could reliably power a home.

Lehman noted that the fuel-cell industry is growing quickly, but that it is rife with secrecy and inflated claims.

Like wind and photovoltaics, hydrogen fuel cells offer certain important advantages over current energy technologies and will no doubt be central features of the post-petroleum infrastructure. We should be dramatically increasing our investments in these alternatives now, while there is still cheap energy to be had. But even assuming a full-scale effort toward a transition to renewables and hydrogen, industrial societies will suffer wrenching changes as a result of the inevitable drastic reduction in available net energy.

Hydroelectricity

While medieval water mills were used to grind grain, modern hydroelectric turbines transform the gravitational potential of rivers and streams into conveniently usable electric power. Electricity generated from water flowing downhill currently constitutes the world's largest renewable energy source.

Throughout the 20th century, hydroelectric dams were built on most major rivers throughout the world — from the Colorado River in the US to the Nile in Egypt. Currently, about 9 percent of electricity in the US is generated by hydro power, a little less than half that generated by nuclear power plants. However, this represents over three times the electricity generated by all other renewable sources combined. In the world as a whole, hydro power accounts for 19 percent of electricity generation.

One of the advantages of generating electricity via hydro dams is that it is relatively easy to store energy during times of low demand. Water empounded behind dams represents stored energy; in addition, surplus electrical power can be used to pump water uphill so that it can be released to flow back through the generating turbines during times of peak demand.

Hydroelectric generation has an attractive EROEI: Odum gives hydro power a net figure of 10, while Cleveland *et al.* assign it 11. Hydro power is thus one of the better current producers of net energy.

Unfortunately, hydroelectric dams typically pose a range of environmental problems: they often ruin streams, cause waterfalls to dry up, and interfere with marine habitat. Dammed rivers are diverted from their geologic and biological work, such as the support of migratory fisheries. Most environmentalists would prefer to remove existing dams rather than see more of them built. Moreover, many existing hydro plants are jeopardized by siltation and foreseeable changes in rainfall patterns resulting from global climate change.

In any case, in the US the building of more large hydroelectric dams is not much of an option. Hydro resources are largely developed; there is little room to increase them. Not one large dam has been approved in the past decade.

The situation is different in Canada, which has immense potential hydroelectric resources. With hydroelectricity as with natural gas, Canada is becoming a major energy source for the US.

Most new hydro developments are being planned not for already-industrialized countries, but for the less-consuming countries of the world. But hydroelectric dams tend to be capital-intensive projects that require huge loans, trapping poor countries in a vicious cycle of debt.

Microhydro — the production of electricity on a small, localized scale from relatively small rivers or streams — offers the advantages of rural electrification with few of the drawbacks of major dam projects. Countless communities in the less-consuming countries may be able to take advantage of this technology, which requires smaller investments and enables local control of resources. Successful microhydro projects are already operating in Sri Lanka, Zimbabwe, the Netherlands, and many other countries.[33] The main drawbacks of such projects are their inability to supply large urban areas with power as well as their reliance on an endangered resource: fresh water.

In sum, hydro power is already a significant energy resource and will continue to be so throughout the coming century. But in many regions of the world — and especially in the US — it is already thoroughly exploited.

Geothermal Power

Humans have enjoyed natural hot springs for millennia, and technologies have more recently been developed for using geothermal waters for home and commercial heating — as is commonly done, for example, in Klamath Falls, Oregon. Underground steam was first used to generate electricity near Rome, Italy, in 1904. The first commercial geothermal electric power plant was built in 1958 in New Zealand; and in 1960, a field of 28 geothermal power plants was completed in the region of Geyserville in northern California.

Geothermal power — whether used for heating or for electricity generation — is necessarily dependent upon geography: plants must be located close to hot springs, geysers, and fumaroles (holes near volcanoes from which vapor escapes). Most geothermal resources are located around the edges of tectonic plates. The west coasts of the Americas as well as Iceland, India, Kenya, the Philippines, Indonesia, Japan, and Thailand all have exploitable geothermal resources.

The US currently has 44 percent of the world's developed geothermal-electric capacity, but the American geothermal industry is stagnant. Less than one percent of the world's electricity production comes from geothermal sources.

By Odum's calculations, geothermal electricity production may currently have an EROEI even higher than that of petroleum (though still far below oil's net yield through the 1960s). However, many geographic locations do not permit the attainment of this degree of net-energy return for geothermal electricity. Moreover, it is unlikely that the generation of electricity from geothermal sources can be increased sufficiently to offset much of the net-energy decline from petroleum depletion.

There is some debate as to whether geothermal electricity production actually constitutes a renewable energy source. As underground steam or hot water is used to turn turbines, it is gradually depleted. The period in which depletion reaches the point where the resource is no longer commercially useful is estimated to be in the range of 40 to 100 years for most geothermal fields. While fields may naturally recharge themselves over a period of centuries or millennia, that will be of little benefit to the next few generations. At The Geysers fields in northern California, efforts are being made to recharge underground reservoirs with treated waste water pumped from the city of Santa Rosa; however, it is too early to tell what the results will be. If successful, the scheme could make geothermal energy production renewable, though the infrastructure and operating costs of the recharging process would drastically reduce the EROEI for energy production from this source.

If recharging efforts fail, the long-term prospects for geothermal electricity look dim. While nations such as Indonesia and Russia have only begun to develop their large potential geothermal resources, without artificial recharging those resources will be useful for only a few decades.

Geothermal energy production has potential for increased local development, but when viewed against the backdrop of the world's total energy needs, its contribution — even if that potential is fully realized — pales in significance.

Tides and Waves

On the shores of oceans, tides rise and fall predictably day by day. This rising and falling of the tides is a potential source of energy. In a few places, estuaries have been dammed so that water can be let in as the tide rises, and then let out via electricity-generating turbines as the tide falls. For an area with 25-foot tides, Odum calculated an EROEI of 15 — which is the highest net-energy yield for any source he studied. However, this net benefit is substantially reduced when the loss of estuarine fisheries is taken into account.

Tidal energy is renewable, clean, and efficient. Unfortunately, there are fewer than two dozen optimal sites for tidal power in the world, and most of those are in remote areas like northwest Russia or Nova Scotia.

The only US city that is likely to benefit significantly from tidal power is San Francisco, which is committed to developing a one-megawatt tidal power station within two years. A major proponent of this project is HydroVenturi Ltd, whose new technology, developed at Imperial College, has no underwater moving parts. As the tide ebbs and flows, long fins inside an underwater passageway would funnel the current, creating suction, which in turn would pull air from pipes connected to onshore turbines, causing the turbines to turn and generate electricity.

If the $2 million test project is a success, it might be possible to power the entire city with electricity generated from the daily tides in the Bay. Potential environmental problems still need to be addressed, including the possibilities that salmon and other fish could be caught in the fins by sudden drops in water pressure; that alteration of the tidal flow could have a negative impact on other marine life; or that increased sediment buildup in the Bay could impair water quality.

Meanwhile a Canadian company, Blue Energy, has created and marketed a highly efficient underwater vertical-axis windmill that can be used to generate tidal power for almost any coastal community. Blue Energy's scalable technology (from a few kilowatts to thousands of megawatts) is claimed to generate efficient, renewable, and emission-free electricity at prices competitive with today's conventional sources of energy. The design of the turbine is structurally and mechanically straightforward, and the transmission and electrical systems are similar to existing hydroelectric installations.[34]

There is also tremendous energy inherent in the waves that constantly lap the ocean shores, and it is theoretically possible to harness some of that energy. But doing so is difficult. Waves are extremely variable: they can occasionally reach 60 feet in height, but days or weeks may go by when the ocean is calm. In Japan, Norway, Denmark, Britain, Belgium, and India, a variety of systems have been used to tap wave energy. The results have been mixed: energy has been produced at relatively low cost, but it tends to be intermittently available. A comprehensive survey of wave-energy research by David Ross suggests that this source can provide only limited power for industrial societies for the foreseeable future.[35]

Biomass, Biodiesel, and Ethanol

"Biomass" is a modern term for what is, in fact, our oldest fuel source: plant material. Current and potential forms of biomass include wood, animal waste, seaweed, peat, agricultural waste such as sugar cane or corn stalks, and garbage.

As noted in Chapter 2, wood was the principal energy source in the US until the latter part of the 19th century, and it still is in many parts of the world. Deforestation in places like Bangladesh and Haiti is directly attributable to the overharvesting of trees for fuel. In the US, biomass provides more total energy than hydroelectric power, making it the nation's principal renewable energy source (though hydro is its foremost renewable source for electricity production).

Biomass has an extremely variable EROEI. However, the burning of all forms of biomass creates air pollution, which can sometimes be severe. Burning wood for heat releases not only carbon dioxide but a cocktail of toxic substances including nitrogen oxides, carbon monoxide, organic gases, and particulate matter. In India, 200 million tons of cow dung are burned annually as cooking fuel; the practice deprives the soil of needed nutrients and also blankets cities in a pollutant haze.

There is limited growth potential for total energy from biomass. Many parts of the world already are experiencing severe and growing shortages of firewood — which is so scarce in parts of Colombia, Peru, India, Pakistan, Bangladesh, Nepal, and some countries of Africa that many people are reduced to having only one cooked meal a day.

In addition to directly burning biomass for heat or light, it is also possible to make fuels from it to run machinery and vehicles. When Rudolf Diesel invented the diesel engine in the late 1890s, he envisioned it running on a variety of fuels, including peanut oil. Today's diesel fuel is a refined petroleum product, but diesel engines can still be modified to run on vegetable oils.

Unmodified diesel engines can burn a fuel known as "biodiesel," which is a chemically altered vegetable oil. The production process for the latter is fairly simple: aside from vegetable oil, the two main ingredients are methanol and lye, and with a little practice and some basic equipment it is possible to produce batches of low-cost biodiesel in one's garage using discarded restaurant deep-fry cooking oil.

Personally, I love biodiesel; I run my car on it. Biodiesel has some distinct advantages over petroleum-based diesel fuel. When burned, it produces fewer pollutants — significantly less CO_2, less particulate matter, no aromatics (benzene, toluemene, xylene), and no sulfur, though nitrogen oxide emissions are the same as with conventional diesel fuel. Mileage per gallon is typically slightly less for biodiesel than for conventional diesel fuel, but users of the former report that the exhaust from their cars or trucks tends pleasantly to smell like French fries or donuts (depending on the oil source).

However, for all its advantages, biodiesel may be destined to remain merely a "boutique" fuel: currently, there are fewer than ten biodiesel plants in the US and only 21 retail pumps scattered throughout the country; moreover, commercial biodiesel sells for over $3 per gallon — significantly more than conventional diesel fuel. An even worse problem is that the production of vegetable oil for use as a fuel is usually, depending on the type of oil, a net energy loser. The National Renewable Energy Laboratory has performed experiments with the extraction of oil from algae, showing that this source could be extremely productive — several times more so than palms or coconuts. However, it has not been shown that this procedure can be scaled up to produce significant commercial quantities of oil. Given the petroleum-intensive nature of modern agriculture, it probably takes more energy to produce a gallon of biodiesel than the biodiesel yields when burned; but if further research on algae oil continues to yield promising results, it is possible that a favorable net-energy production could be achieved and a sizeable portion of the diesel fleet could be run on biofuels.

While most enthusiasts use vegetable oil in the form of biodiesel, some modify their diesel car's fuel system to accept ordinary, recycled vegetable oil. Both strategies appeal to a tiny but growing number of environmentally aware motorists who have started fuel-sharing co-ops and who maintain websites devoted to the promotion of vegetable oil-fueled transportation. While there simply aren't enough fast-food restaurants or donut shops to fuel large fleets of cars and trucks, this is a good option for the few mavericks willing to make the effort.

Ethanol — a fuel-grade form of alcohol produced from grain fermentation — suffers from the net-energy constraints similar to those of biodiesel. Promoters tout ethanol as a clean energy alternative since it produces fewer pollutants when burned than do petroleum byproducts, and the US Congress has adopted laws requiring ethanol to be mixed with gasoline for automobile consumption. Essentially, this Federal mandate amounts to a subsidy for agribusiness, since ethanol is produced primarily from corn grown in the American Midwest. Altogether, the ethanol industry receives about $1.4 billion per year in direct or indirect subsidies, most of which end up benefiting giant agribusiness cartels such as Archer Daniels Midland.

Cornell University professor David Pimentel, who has performed two net-energy analyses of ethanol, found in both instances that the fuel cost more energy to produce than it eventually delivered to society. While his recent study was more favorable than the previous one, it nevertheless showed an EROEI of roughly 0.81, meaning a 29 percent net loss of energy.[36]

However, Pimentel's studies have been attacked by ethanol proponents, who cite much more favorable reports — especially several USDA studies led by Hosein Shapouri, the most recent of which comes to the optimistic conclusion that ethanol offers up to a 77 percent energy profit.[37]

But Shapouri's and other ethanol-favorable studies have in turn been devastatingly critiqued by Tad W. Patzek of University of California, Berkeley, in a 114 page paper titled "Thermodynamics of Corn-Ethanol Biofuel Cycle."[38] Patzek argues that Shapouri has disregarded or minimized several important energy inputs to the process of ethanol production; once these figures are corrected, the net energy gain cited in the USDA studies is "insupportable."

Ethanol proponents correctly point out that crops other than corn (such as sugar cane) can yield more alcohol per acre; moreover, engines that burn ethanol may last considerably longer than gasoline-burning engines, thus leading to energy savings elsewhere in the industrial system.

Proponents also point to Brazil's experimental use of ethanol from sugar cane as a vehicle fuel in the 1980s. An impressive 91 percent of Brazilian cars produced in 1985 ran on sugar-cane ethanol. However, as world oil prices plummeted during the latter half of the decade, and as sugar prices rose, demand for alcohol-fueled cars subsided. It could be argued that Brazil was able to afford its ethanol experiment primarily because of its favorable ratio of available cropland to automobiles: even if energy and topsoil were being lost in the exercise, the country was temporarily able to absorb these losses because they were small and temporary; the situation would likely be very different in the US.

Brazil remains the world's largest producer of ethyl alcohol, supplying 38 percent of the worldwide total. Yet many environmentalists have expressed fears that if demand for ethanol accelerates, Brazil could be transformed into one giant sugar cane field. Already Brazilian agriculture is encroaching on the *cerrado,* a vast grassland and savannah region in the southeastern section of the central Brazilian plateau constituting a unique and seriously threatened ecosystem.

If the US were to attempt to imitate Brazil's feat, how much farmland would be needed to provide enough ethanol to replace fossil fuels? The United States has about 400 million acres of cropland and about 200 million cars. American farmers produce about 7,110 pounds of corn per acre per year, and an acre of corn yields about 341 gallons of ethanol. The typical American driver would burn 852 gallons of ethanol per year, thus requiring 2.5 acres of cropland. According to this calculation, 500 million acres of farmland would be needed to provide fuel for the American fleet — or 25 percent more farmland than currently exists. (This assumes that no farmland would be used to grow food.) While ethanol advocates point out that land used for ethanol production can simultaneously be used to produce cattle feed (which is how corn is mostly used these days anyway), the above calculation should nevertheless give us pause, especially given the fact that in 2005 the US will become a net food importer (in dollar terms) for the first time in its history as a nation.[39]

Even if we accept the controversial claim that ethanol can be produced in such a way as to yield a net energy profit, it would be foolish to assume that a large percentage of US fleet could be run on the fuel, given the above environmental constraints. If our goal is a sustainable energy regime, it is more realistic merely to envision organic farmers devoting a portion of their land to the production of modest amounts of ethanol with which to run their farm machinery.

Fusion, Cold Fusion, and Free-Energy Devices

Some people maintain that there are energy sources not constrained by the laws of physics as presently understood. It would be simple enough to write off this viewpoint as pseudoscience; however, in the context of the resource depletion discussion, such claims deserve to be addressed. Are free-energy devices possible?

In essence, a free-energy (or "over-unity") device is one that produces more power than it consumes in its operation. The search for free energy (formerly referred to as "perpetual motion") began long ago. In the 14th century, Villand de Honnecourt produced a drawing of a perpetual-motion machine, as did Leonardo da Vinci a couple of centuries later. Johanes Taisnerius, a Jesuit priest, worked on a perpetual-motion machine based on permanent magnets; and Cornelis Drebbel, an alchemist and magician, supposedly made one in 1610. The first English perpetual-motion patent was granted in 1635; by 1903, 600 such patents had been granted. In the 19th century, so many people were working on *perpetual-motion* machines that their goal inspired a musical genre — the perpetuum mobile — which transfixed the audiences of virtuosi like Nicolo Paganini and Carl Maria von Weber.

In the 20th century, the free-energy literature tended to center on the work of Nikola Tesla (whose career is briefly discussed in Chapter 2). Tesla produced most of his useful inventions before 1910; thereafter his work became progressively more obscure — some would say, visionary. According to one often repeated (and likely apocryphal) anecdote, in 1931 the reclusive inventor fitted a new Pierce-Arrow with a mysterious 80-horsepower, alternating-current electric motor that had no batteries and no external power source, and drove it for a week.

Unlike Tesla, most 20th-century claimants to perpetual motion were relatively obscure figures. In the 1920s, a self-taught inventor named Lester Hendershot built a generator comprising twin basket-weave coils, capacitors, transformers, and an input magnet/clapper unit, which reputedly produced useful electrical power at about 300 watts. The device tended to be erratic, as its operation depended on the tuning of the input component; moreover, Hendershot himself was unable to provide a scientific explanation of how the device worked.

Also in the 1920s, Dr. T. Henry Moray of Salt Lake City began experimenting with solid-state circuitry, cold cathode-ray tubes, and a radiant-energy device that produced up to 50 kilowatts of electrical power. Similar radiant-energy devices were developed independently by L. R. Crump, Peter Markovitch, and others. Several patents were granted, and efforts have more recently been made to explain the phenomenon in terms of "neutrino flux" and "tachyon fields."

In the 1980s, an inventor named Joseph Newman introduced a series of machine generators built around a powerful permanent magnet rotating within a coil consisting of a very large number of turns of copper wire. In the course of his promotional efforts, Newman gave a weeklong demonstration in the Super Dome in New Orleans and appeared on the *Tonight* show. He claimed that his machine produced much more energy than it consumed, but critics maintained that the apparent surplus of power (most of which was dissipated in heat) was actually the result of measurement errors.

The above in no way constitutes an exhaustive list of perpetual-motion or free-energy claimants. There are and have been literally scores of others — some deliberate hoaxers, others sincere but naïve backyard tinkerers, and still others serious scientists. Many of their efforts seem transparently quixotic. Their goal is clear enough: If only we can find a new, infinite source of energy, we can free ourselves from all sorts of material constraints. But how do free-energy researchers explain — to themselves or their investors — that such a thing is even possible?

The standard textbook view of energy begins with the First Law of Thermodynamics, which states that the sum of all matter and energy in the universe is constant and that energy can be neither created nor destroyed — only its form changes. There are no exceptions: this is the most fundamental law by which we must live, one that cannot be bent, broken, or repealed. What makes free-energy advocates think they can get around it? Is there a loophole?

The best chance of finding one, some suggest, is by way of quantum physics. Theoretical physicists speculate that empty space may not really be empty after all; it may, in fact, be filled with energy. If so, all we would need to do to harvest that energy would be to assemble the equivalent of a quantum windmill to capture the quantum "wind."

If the details of the process are a bit abstruse, the fact that the well-known science fiction writer Arthur C. Clarke has endorsed the possibility of obtaining energy from vacuum — this is sometimes called "zero-point" energy — is encouraging. Perhaps the search for new energy sources has outgrown the garages of inventors like Hendershot and is ready to move into university physics labs.

Another potential path toward free energy is cold fusion. In 1989, physicists Stanley Pons and Martin Fleishman of Salt Lake City announced that they had produced a nuclear-fusion reaction at room temperature — a feat previously considered impossible. Cold fusion reputedly occurs when ordinary hydrogen and an isotope of hydrogen called deuterium are brought together with metals such as palladium, titanium, and lithium. The reaction (again, reputedly) releases enormous quantities of energy — more than ordinary chemical reactions could possibly yield. Cold fusion, in contrast to hot fusion, happens in a relatively simple apparatus roughly the size of a postage stamp and does not emit neutron radiation. It also gives off very little, if any, of the radiation common to nuclear-fission reactions.

Many American scientists still consider cold fusion a form of crank science, though well over 1,000 peer-reviewed papers on the subject have been published. Cold-fusion researchers have never claimed that the effect produces power from vacuum or that it violates any known laws of physics.[40]

What impact will any of these efforts to develop exotic energy devices have on the energy shortages of the 21st century? In the near term, very little. In all likelihood, most if not all of the ballyhooed free-energy claims of the past were the result of deliberate deception, measurement error, or naïveté on the part of unschooled researchers. Moreover, it is difficult to avoid the impression that many of the current Internet discussions of exotic energy devices are pervaded by paranoia and extravagant claims, such as, "The oil companies are buying up all the patents and suppressing the evidence!" or "A secret, unelected military government is running free-energy 'black' projects with technology stolen from space aliens!" or, "Our only hope is to quickly fund this or that maverick inventor, whose latest device generates a million times more power than it consumes!"

Is the US government really secretly experimenting with free-energy devices? That is entirely possible. It is even possible (in the sense that almost anything is possible) that the technology was acquired from space aliens. The problem with discussing the subject is that most secret government programs are surrounded with disinformation spread by well-paid experts. Given the continual rain of lies and half-truths about military or intelligence "black" projects, it is impossible to know what to believe about them, and under such circumstances most speculation is a waste of time.

Sensationalism aside, it appears that, even if cold-fusion devices or "quantum windmills" could work, harnessing these new power sources would not be easy. Currently, the world derives exactly zero percent of its commercially produced energy from all of these exotic sources combined. It is likely that, even in the best case, decades of further research and development would be required to change that statistic appreciably.

Energy production from conventional or hot nuclear fusion is less controversial from a theoretical point of view than are proposed zero-point or cold-fusion projects. Billions of research dollars have been devoted to fusion research over the past two decades. If made practical, fusion could produce almost limitless energy from seawater. However, the hurdles to actually producing fusion energy are prodigious. Reactor temperatures would have to be in the range of 360 million degrees Fahrenheit (200 million degrees Celsius), and no materials or processes are currently capable of containing such temperatures for more than a tiny fraction of a second. No fusion reactor has yet succeeded in producing more energy than it consumes. Even promoters say that commercially useful power production from fusion is at least 50 years away — but it may not be possible to continue funding expensive and energy-intensive fusion research in the energy-constrained environment of the 21st century.

Conservation: Efficiency and Curtailment

Nearly everyone agrees that the best ways to cushion the impact of an energy shortage are simply to consume less and to get the most out of what we do use. The term *conservation* is often employed to refer to these two parallel but fundamentally different strategies. The first strategy — perhaps more accurately termed *curtailment* — would, for example, translate into the prosaic action of turning off a light when leaving a room. The second — more accurately termed *efficiency* — would, in terms of the same example, mean replacing an incandescent bulb with a compact fluorescent bulb that produces the same amount of light from a quarter of the electricity. There is plenty of room for energy savings from both strategies.

In the past three decades, American homes and workplaces have become much more energy efficient. In the 1950s, the US economy as a whole used over 20,000 BTU for every inflation-adjusted dollar of gross domestic product; by 2000, it was consuming only about 12,000 BTU per dollar. Much of this improvement in efficiency was due to the redesign of common appliances such as refrigerators, lamps, and washing machines. Today's houses are typically insulated better than houses a few decades ago, and most buildings and factories have been redesigned for energy efficiency.

More such gains are possible. Between 1980 and 1995 the fuel efficiency of US automobiles improved dramatically, but since then that trend has reversed. Cars themselves did not become less efficient; instead, many drivers — encouraged by low gas prices — began buying light trucks or sport utility vehicles, which typically use much more fuel than smaller cars. Toyota and Honda have begun marketing hybrid gasoline-electric cars that achieve over 50 miles per gallon, and American automakers are beginning to roll out their own hybrid versions of existing cars — including SUVs. In the future, an 80 mpg full-size car is probably feasible. Many homes can still benefit greatly from extra insulation, low-e windows, the planting of shade trees to reduce the need for air conditioning, and the replacement of incandescent bulbs with compact fluorescents.

Substantially increased energy savings from efficiency are also possible in industry. Philips, a large European manufacturing firm, is a typical success story in this regard. After deciding in the early 1990s to target energy efficiency, the company hired consultants and began making changes in its operations. Between 1994 and 1999, Philips improved energy efficiency by 31 percent, while reducing its waste stream by 56 percent.

The efficiency of US electricity generation plants peaked in 1958 at about 35 percent. However, newer plant designs are able to achieve efficiencies of 57 percent or more. In 1998, two-thirds of electric generating plants were more than 25 years old; replacing half of these with new, more efficient plants could increase available electricity by about 25 percent with no increase in fossil fuel consumption. Moreover, waste heat from generating plants could be employed to heat homes and factories, or to raise the efficiency of hydrogen production.

Many of these potential improvements could be speeded up through offering subsidies or tax incentives, and energy markets could benefit greatly from intelligent regulations that promote efficient energy provision and consumption. Such an intelligent redesign of regulations in the UK in the 1990s led to a significant increase in energy efficiency, a decrease in the use of nuclear power, and a 39-percent reduction in CO_2/kWh.

However, there are limits to the benefits from efficiency, since increasing investments in energy efficiency typically yield diminishing returns. Initial improvements tend to be easy and cheap; later ones are more costly. Also, the energy costs of retooling or replacing equipment and infrastructure can sometimes wipe out gains. A simple example: Suppose you are currently driving a two-year-old car that travels 25 miles on a gallon of gasoline. You see a similar new car advertised that gets 30 mpg It would appear that, by trading cars, you would be conserving energy. However, the situation is not that simple, since a little over ten percent of all the energy consumption attributable to each vehicle on the road occurs in the manufacturing process — before that vehicle has traveled its first mile. Thus, by putting off trading cars you might be conserving more net energy than you would be by buying the new, more fuel-efficient replacement.

In the late 1980s, Gever *et al.* studied the relationship between energy efficiency and national economies, as expressed in the ratio between gross domestic product (GDP) and total energy consumed (a rising ratio of GDP to energy consumption means that the economy is becoming more energy-efficient). Not surprisingly, they found that nations like Sweden, Switzerland, and Denmark were much more energy-efficient than the US, and that US energy efficiency had improved significantly during the 1980s. But what were the factors driving increased efficiency? Their analysis showed that energy efficiency increases with the use of more energy-dense sources — this is by far the most important factor — as well as with the reduction of household use of energy and with increased energy prices. As industrial nations made the transition from burning coal to using higher net-yield sources — oil, gas, hydro, and nuclear power — energy efficiency improved dramatically. Household energy consumption (which goes mostly to heating homes and fueling cars) does not add as much to the GDP as does industrial use of energy, which goes toward the production of goods and services, and so efficiency improved as households used proportionately less. And higher energy prices encouraged the switch to more energy-stingy technologies. But the authors pointed out that:

> our analysis indicates that the ability of technical change to increase the goods and services produced from the same amount and mix of fuels is much smaller than most economists claim .... There are several reasons to believe that previous assessments of technology's ability to save energy were overly optimistic. For one, many analyses ignored important changes in the kinds of fuels used in the economy and in the division of fuel supplies between household and intermediate sectors. As a result, changes in efficiency due to these factors were mistakenly attributed to technological advances and/or fuel prices ....[41]

The authors also noted that:

> [i]n agriculture, for example, the amount of fuel used directly on a cornfield to grow a kilogram of corn fell 14.6 percent between 1959 and 1970. However, when the calculation includes the fuel used elsewhere in the economy to build the tractors, make the fertilizers and pesticides, and so on, it turns out that the total energy cost of a kilogram of corn actually rose by 3 percent during that period.[42]

The inescapable implications of these findings are first, that many efforts toward energy efficiency actually constitute a kind of shell game in which direct fuel uses are replaced by indirect ones, usually in the forms of labor and capital, which exact energy costs elsewhere; and second, that the principal factor that enabled industrial countries to increase their energy efficiency in the past few decades — the switch to energy sources of higher net yield — does not constitute a strategy that can be applied indefinitely in the future.

Thus the curtailment of energy usage offers clearer benefits than improved efficiency. By simply driving fewer miles one unequivocally saves energy — regardless of whether one's car is old or new and whether it is more or less efficient.

Some curtailment is painless — as is the case with turning off the lights when one leaves a room or turning down the thermostat at night. But the economy as a whole is inextricably tied to energy usage, and so significant degrees of curtailment throughout society are likely to have noticeable economic consequences.

We have historical data in this regard. In the 1970s and early 1980s, the US curtailed some of its energy usage due to the oil-price shocks of 1973 and

1979. People drove fewer miles in smaller cars and drove more slowly due to lowered speed limits. As a result, the national GDP/energy ratio improved — but at some cost in terms of the standard of living. That cost was relatively easily borne, but that it was indeed a cost is shown by the fact that when fuel prices drifted back downward, people again began driving more and faster, and choosing larger cars.

Given that, from a historical and cross-cultural perspective, Americans' average standard of living is lavish, it would seem that some curtailment of consumption may not be such a bad thing. After all, people currently have to be coaxed and cajoled from cradle to grave by expensive advertising to consume as much as they do. If the message of this incessant propaganda stream were simply reversed, people could probably be persuaded to happily make do with less. Many social scientists claim that our consumptive lifestyle damages communities, families, and individual self-esteem; a national or global ethic of conservation could thus be socially therapeutic.[43]

However, eventually curtailment means reducing economic activity — it means fewer jobs, goods, and services. It means fundamental changes not only in the *pattern* of life but also in the *quality* of life that we have become accustomed to. Mild degrees of curtailment in national energy usage might just involve sacrifices of speed and convenience. Intermediate degrees might imply tradeoffs in health care, transportation, housing space, and entertainment options. But severe curtailment — unless undertaken systematically over a period of decades — would likely lead to rampant unemployment and shortages of basic necessities.

Energy conservation — both increased efficiency and curtailment of energy usage — will be crucial in cushioning impacts from the depletion of oil. But it is not a panacea.

With such a broad array of alternatives to choose from, many people assume it must be possible to cobble together a complex strategy to enable a relatively painless transition away from fossil fuels. Surely, for example, by building more wind turbines and fuel cells, by exploiting advances in photovoltaic technologies, and by redoubling our national conservation efforts, we could effortlessly weather the downside of the Hubbert curve.

A recurring subtext of this chapter has been the importance of net-energy analysis. To date, very few such analyses have been performed by impartial and competent parties. It is essential to the welfare of current and future generations that a standardized and well-defined net-energy methodology be adopted by national and international planning agencies. Reliance on market price as a basis for energy policy is shortsighted, because hidden subsidies so often distort the picture. Any standardized EROEI evaluation methodology will inevitably be imperfect, but it will nevertheless provide the public and decision makers alike with much sounder insights into the costs of various energy options before precious resources are committed to them. As we have seen, the net-energy returns for some renewables (particularly wind) already exceed the dwindling returns for nonrenewable coal and domestic petroleum. Other options (such as hydrogen) may lose their luster when looked at closely.

Clearly, we would see the best outcome if all of the nations of the world were to undertake a full-scale effort toward conservation and the transition to renewables, beginning immediately. And undoubtedly some sort of complex strategy will eventually be adopted. But we should not delude ourselves. Any strategy of transition will be costly — in terms of dollars, energy, and/or our standard of living. Odum and Odum summarize the situation succinctly: "Although many energy substitutions and conservation measures are possible, none in sight now have the quantity and quality to substitute for the rich fossil fuels to support the high levels of structure and process of our current civilization."[44]

This is somewhat of a double message. Renewable alternatives are capable of providing net-energy benefit to industrial societies. We *should* be investing in them and converting our infrastructure to use them. If there is any solution to industrial societies' approaching energy crises, renewables plus conservation will provide it. Yet in order to achieve a transition from nonrenewables to renewables, decades will be required — and we do not have decades before the peaks in the extraction rates of oil and natural gas occur. Moreover, even in the best case, the transition will require shifting investment from other sectors of the economy (such as the military) toward energy research, conservation, and the implementation of renewable alternatives. Those alternatives will be unable to support the *kinds* of transportation, food, and dwelling infrastructure we now have; thus the transition will necessarily be comprehensive: it will entail an almost complete redesign of industrial societies. The result — an energy-conserving society that is less mobile, more localized, and more materially modest — may bring highly desirable lifestyle benefits for our descendants. Yet it is misleading to think that we can achieve that result easily or painlessly.

If indeed none of the energy alternatives now available has the near-term potential to "support the high levels of structure and process of our current civilization," then profound changes are virtually inevitable in every sphere of human concern as oil begins to run out. Just what sorts of changes can we expect to see within the next 50 years?

5

A Banquet of Consequences

Anyone who believes exponential growth can go on forever in a finite world is either a madman or an economist.

— Kenneth Boulding (ca. 1980)

If we continue ... to consume the world until there's no more to consume, then there's going to come a day, sure as hell, when our children or their children or their children's children are going to look back on us — on you and me — and say to themselves, "My God, what kind of monsters were these people?"

— Daniel Quinn (2000)

Current debates over where and how to drill for oil in this country soon may be rendered irrelevant by a nation desperate to maintain its quality of life and economic productivity. War over access to the diminishing supply of oil may be inevitable unless the United States and other countries act now to develop alternatives to their dependence on oil.

— Senator Mark Hatfield (1990)

We need an energy bill that encourages consumption.

— George W. Bush (2002)

Sooner or later, we sit down to a banquet of consequences.

— Robert Louis Stevenson (ca. 1885)

When the global peak in oil production is reached, there will still be plenty of petroleum in the ground — as much that will be ultimately recoverable as has been extracted from 1859 to the present, or roughly one trillion barrels (by most estimates). But every year from then on, it will be difficult or impossible to find and pump as much oil as the year before. The rate-curve of extraction necessarily roughly mirrors, with a time lag, the rate-curve of discovery — which peaked in 1964. Even if efforts are intensified now to switch to other energy sources, those efforts will come so late that, for the duration of the transition, society will inevitably have less net energy available to do useful work — including the manufacturing and transporting of goods, the growing of food, and the heating of homes.

During the past two centuries, we have become accustomed to a regime in which there was *more* energy available each year, and our population has grown quickly to take advantage of this energy windfall. We have come to rely on an economic system built on the assumption that growth is normal and necessary, and that it can go on forever.

As we move from a historic interval of net-energy growth to one of net-energy decline, we are entering uncharted territory. It takes some effort to adjust one's mental frame of reference to this new reality.

Try the following thought experiment. Go to the center of a city and find a comfortable place to sit. Look around and ask yourself: Where and how is energy being used? What forms of energy are being consumed, and what work is that energy doing? Notice the details of buildings, cars, buses, streetlights, and so on; notice also the activities of the people around you. What kinds of occupations do these people have, and how do they use energy in their work? Try to follow some of the strands of the web of relationships between energy, jobs, water, food, heating, construction, goods distribution, transportation, and maintenance that together keep the city thriving.

After you have spent at least 20 minutes appreciating energy's role in the life of this city, imagine what the scene you are viewing would look like if there were 10 percent less energy available. What substitutions would be necessary? What choices would people make? What work would not get done?

Now imagine the scene with 25 percent less energy available; with 50 percent less; with 75 percent less.

Assuming that the peak in global oil production occurs at some point in the period from 2006 to 2010 and that there is an average two percent decline in available net energy each year afterward, in your imagination you will have taken a trip into the future, to perhaps the year 2050.

This exercise is, of course, speculative. However, some speculation — if grounded in an awareness of present reality and existing trends — can be helpful.

This chapter consists of speculations about the effects of the net-energy decline during the next few decades. Forecasts are always fallible, as anyone who regularly reads five-day weather forecasts knows. But some scenarios of future events are more likely than others; and with an understanding of how epochs of energy abundance or shortage have shaped the development of societies in the past, we should be able to foresee some of the general outlines of events to come as industrial societies move from a condition of energy abundance to one of energy scarcity.

Clearly, the energy transition of the early 21st century will affect nearly everything that humans care about. No person or group will be untouched by this great watershed.

Because the shift will be incremental, it would be a mistake to assume that the effects discussed in this chapter will all occur soon or in an instantaneous fashion. However, it would also be a mistake to assume that they will be so gradual in their appearance that they will accumulate to truly dramatic proportions only in our grandchildren's lifetimes or later. The early effects of the net-energy peak are already upon us and will probably begin to cascade within the next two decades or even the next few years.

Industrial civilization is a complexly interrelated entity, and it will respond to the net-energy decline as a system. It is therefore problematic to deal separately with effects on agriculture, transportation, and economics because developments in any one area will impact — and be impacted by — developments elsewhere in the system. However, written information must necessarily be organized in a linear fashion, so we will deal with various aspects of society one by one. As we do so, the reader may wish to give some thought to the ways in which each aspect dovetails with the others.

The Economy — Physical and Financial

The links between the physical economies of nations — their production, distribution, and consumption of goods and services — and the availability of energy are fairly obvious, but the subject bears some discussion nevertheless. All human activities require energy, which physicists define as the capacity to do work. With less net energy available, less work can be done — unless the efficiency of the process of converting energy to work is raised at the same rate as that at which energy availability declines. It will therefore be essential, over the next few decades, for all economic processes to be made as energy-efficient as possible. However, as discussed in the previous chapter, efforts to improve efficiency are subject to diminishing returns, and so eventually a point will be reached when reduced energy availability will translate into reduced economic activity.

Our current financial system was designed during a period of consistent growth in available energy, with its designers operating under the assumption that continued economic growth was both inevitable and desirable. This *ideology* of growth has become embodied in systemic financial structures *requiring* growth. The most prominent of these is compound interest.

Suppose you were to deposit $100 in a bank account earning six percent interest, and left it there for your children or grandchildren. After the first year, you would have $106, and after the second, $112.30. In twelve years your deposit would have doubled, and in a hundred years it would grow to $33,930. Unfortunately, the compound interest on debt works the same way: if you were to take out a loan of $100 at six percent interest and fail to make payments, over time that debt would grow similarly.

Currently all nations have a type of monetary system in which virtually all money is created through the making of loans. Thus, nearly all of the money in existence represents debt. For those not familiar with banking, this may be a difficult fact to grasp: I find that when I present it to college students, I often have to reiterate it in various ways for an hour or so before they are able to comprehend that money is not a physical substance kept in a vault, but a fictitious entity created out of nothing by bankers in order to facilitate the keeping of accounts.

All of this being so, a problem arises: From where does the money come with which to pay back the *interest* on loans? Ultimately, that money has to come from new loans, taken out by others somewhere else within the financial network of the economy. If new loans are not being made, then somewhere in the network people will be finding it impossible to pay the interest on their existing loans, and bankruptcies will follow. Thus the necessity for growth in the money supply is a structural feature of the financial system. The system seems to function best when growth in the money supply is kept at a low and fairly constant rate, and this is the job of the national banks (in the US, the Federal Reserve; in Canada, the Bank of Canada; in England, the Bank of England, and so on), which adjust interest rates to this end.

If money creation (i.e., the making of loans) occurs more rapidly than the growth in the production and consumption of goods and services in the economy, then *inflation* results; money then has less purchasing power, and this is bad for lenders — since the money used to repay loans is then worth less than the money that was borrowed. If money is not being loaned out (i.e., created) at a pace fast enough to match the growth in goods and services, then not enough money will be available to repay existing loans (plus interest), and the resulting bankruptcies and foreclosures can — in extreme cases — cause the economy to go into a tailspin of cascading financial cannibalism. *Deflation* may ensue, in which the purchasing power of money actually increases.

Until now, this loose linkage between a financial system predicated upon the perpetual growth of the money supply and an economy growing year by year because of an increasing availability of energy and other resources has worked reasonably well — with a few notable exceptions, such as the Great Depression. Productivity — the output produced per worker-hour — has grown dramatically, not because workers have worked harder but because workers have been controlling ever more energy in order to accomplish their tasks. Productivity, total economic activity, population, and money supply have all grown — at rates that have fluctuated, but within acceptable ranges.

The lower-energy economy of the future will be characterized by lowered productivity. There could be a good side to this in that more human labor will be required in order to do the same amount of work, with human muscle-power partially replacing the power of fossil fuels. Theoretically, this could translate into near-zero unemployment rates.

However, the financial system may not respond rationally. With less physical economic activity occurring, businesses would be motivated to take out fewer loans. This might predictably trigger a financial crisis, which would in turn likely undermine any attempts at a smooth economic adjustment.

As Hubbert pointed out, the linkage between the money system (the financial economy) and the human matter-energy system (the physical economy) is imperfect. It is possible for a crisis to occur in the financial system even when energy, raw materials, and labor remain abundant, as happened in the 1930s. But is it also possible for the financial system to remain healthy through an energy-led decline in the physical economy? That, unfortunately, is highly unlikely, due to the dependence of the former on continued borrowing to finance activity in the latter. Rather, it is highly likely that the net-energy decline will sooner or later trigger a financial crisis through a reduction in demand for goods and services, and hence for money (via loans) with which to pay for the machinery to produce those goods and services. Thus even if human labor is sufficiently abundant to make up for some of the reduction of work performed by fuel-burning machines, the financial system may not be able to adapt quickly enough to provide employment for potential laborers.

Therefore extreme dislocations in both the financial system and the human matter-energy economy are likely during the energy transition. The exact form these dislocations will take is difficult to foresee. Efforts could be made to artificially pump up the financial system through government borrowing — perhaps to finance military adventures. Such massive, inflationary borrowing might flood markets with money that would be losing its value so quickly as to become nearly worthless. On the other hand, if inflationary efforts are not undertaken quickly or strenuously enough when needed, then the flagging rate of loans might cause money to disappear from the economy; in that case, catastrophic deflation would result. As was true in the Great Depression, what little money was available would have high purchasing power, but there would simply be too little of it to go around. Unemployment, resource and product shortages, bankruptcies, bank failures, and mortgage foreclosures would proliferate.

It is entirely possible that, over a period of decades, both inflationary and deflationary episodes may occur; however, due to the lack of a stable linkage between money and energy, periods of financial stability will likely be rare and brief.

Continued population growth, even at reduced rates, will put added strain on support systems and exacerbate the existing inherent requirement for economic growth.

Who will feel the pain? Most likely, the poor will feel it first and hardest. This will probably be true both nationally and internationally, as rich nations will likely seek to obtain energy resources from the poorer nations that have them by financial chicanery or outright military seizure. Eventually, however, everyone will be affected.

Some comforts, even luxuries, will probably continue to be available in most countries; but regardless of whether the financial environment is inflationary or deflationary, nearly everything that is genuinely useful will become relatively more expensive because the energy employed in its extraction or production will have grown more rare and valuable.

Transportation

The automobile is one of the most energy-intensive modes of transportation ever invented. This is true not just because of its direct use of fuel (a lightly loaded bus, airliner, or train actually uses more fuel per passenger-mile) but for the energy embodied in the construction of so many individual units that require replacement every few years. The rate of car ownership in the US is now 775 per thousand people — nearly the highest in the world — and many less-consuming nations, such as China, are foolishly seeking to emulate the American love affair with the automobile. Because increased car ownership results in changed patterns of urban development and resource distribution, it creates social dependency. Wherever this dependency has taken hold, it will have ruinous consequences in the coming century.

Over the short term, more energy-efficient cars will be built, including gasoline-electric hybrids, and perhaps some hydrogen-powered models. But the relentless economics of the energy decline will mean that — eventually but inevitably — fewer cars will be built. Only the wealthy will be able to afford them. The global fleet of autos will gradually age and diminish in number through attrition. For a peek at the year 2050, look to Cuba — where 50-year-old Fords and Plymouths are still in service because virtually no newer ones have been imported from the US due to the trade embargo.

During the 20th century, millions of miles of roads and highways were built for automobile and truck traffic, at extraordinary expense. The Los Angeles Freeway, for example, cost taxpayers $127 million per mile to construct. In fiscal 1995 alone, local, state, and federal governments in the US spent $80 billion on roads and highways. Last century's prodigious road-building feat — dwarfing any of the wonders of the ancient world — was only possible because oil was cheap. Asphalt incorporates large quantities of oil, and road-building machines run on refined petroleum. In the decades ahead, road building will grind to a halt and existing roads will gradually disintegrate as even repair efforts become unaffordable.

Countries with good public transportation — street cars, buses, subways, and trains — will be much better poised than the US to weather the energy transition. Mass-transit users typically spend $200 to $2,000 per year for travel, considerably less than car owners spend. Also, when well utilized, mass-transit systems consume much less energy per passenger mile than do automobiles. In her book *Divorce Your Car! Ending the Love Affair with the Automobile,* Katie Alvord points out that "[w]hile a single automobile uses over 5,000 BTUs per passenger mile, a train car carrying 19 people uses about 2,300 and a bus carrying the same number only about 1,000."[1] However, the construction of mass-transit systems itself requires a sizable energy investment, and the US — where the development or maintenance of mass transit was actively discouraged in favor of the private automobile — will find it increasingly difficult to make such investments.

Modern passenger jets run on high-grade kerosene refined from oil. The only likely replacement fuels are ethanol and hydrogen, which would offer some advantages but also pose serious problems.

Ethanol produced from biomass would be expensive to produce in quantity and would require the redesign of jet engines; existing propeller-driven aircraft could burn ethanol more readily. However, as discussed in the previous chapter, the net energy gain from ethanol production is at best minimal, and the amount that could be produced is limited by the amount of available cropland.

Hydrogen contains three times as much energy per unit of weight as does kerosene, which means that only one-third as much fuel would have to be carried by hydrogen-burning planes to cover a similar range. However, hydrogen's lower density requires the complete redesign of aeronautic fuel tanks. In order to reduce the space needed for hydrogen storage, the fuel would need to be kept in liquid form at minus 285 degrees Fahrenheit (minus 253 degrees Celsius). But even then the specific volume of hydrogen is twelve times greater than that of kerosene, so hydrogen storage tanks would necessarily be much larger than those for kerosene. Moreover, with such low temperatures being maintained on board the aircraft, considerable insulation would be required, which would add to aircraft weight. Other technical problems include the requirement for a redesign of aircraft engines to properly burn the alternative fuel.

NASA experimented with hydrogen-powered aircraft in the 1950s and 1960s, when a B-57 jet bomber flew partially on liquid hydrogen. The former Soviet Union also tried hydrogen experimentally, converting one of the three engines of a Tupolev 154 passenger jet to liquid hydrogen. Currently, NASA is supporting new research on hydrogen-powered aircraft. But as of today there are no commercial airliners that run on hydrogen, nor are any likely to be built for at least two decades.

As we saw in the previous chapter, the production of hydrogen in large quantities presents problems — both from the standpoint of the need for natural gas as a feedstock and because of the low net-energy yield for most of the electricity sources that could produce hydrogen from water through electrolysis.

There is thus no doubt that, whether it depends on kerosene, ethanol, or hydrogen, air travel will become extremely expensive as the 21st century wears on. Given that oil will still be available throughout most of the coming century, though at much higher prices, it is possible that rich individuals will continue to avail themselves of air travel in some form and that the military will increasingly commandeer dwindling flight fuels for fighters, bombers, helicopters, and missiles. But it is highly unlikely that the commercial airline industry as we know it today will survive any attempted transition to ethanol or hydrogen. As a result, the tourism industry will languish in the decades ahead. This could have devastating effects on places like Hawaii, whose economies are almost entirely dependent on tourism.

But even more serious consequences of reduced transportation will be felt in disruptions in the distribution of goods. In the 1980s and '90s, increased global trade resulted in the moving of products and raw materials ever further distances from source to end user. As transportation fuels dwindle — for air, sea, and land travel — we will see an inevitable return to local production for local consumption. But this process of "globalization in reverse" will not be without difficulty, since local production infrastructures were often cannibalized in the building of the global economy. For example, no large shoe companies continue to manufacture their products in the US. Unfortunately, the rebuilding of local production infrastructures will be problematic with less energy available.

Food and Agriculture

Throughout the 20th century, food production expanded dramatically in country after country, and virtually all of this increase was directly or indirectly attributable to energy inputs. Since 1940, the productivity of US farmland has grown at an average rate of two percent per year — roughly the same pace as that by which oil consumption has increased. Overall, global food production approximately tripled during the 20th century, just keeping pace with population growth.

Modern industrial agriculture has become energy-intensive in every respect. Tractors and other farm machinery burn diesel fuel or gasoline; nitrogen fertilizers are produced from natural gas; pesticides and herbicides are synthesized from oil; seeds, chemicals, and crops are transported long distances by truck; and foods are often cooked with natural gas and packaged in oil-derived plastics before reaching the consumer. If food-production efficiency is measured by the ratio between the amount of energy input required to produce a given amount of food and the energy contained in that food, then industrial agriculture is by far the least efficient form of food production ever practiced. Traditional forms of agriculture produced a small solar-energy surplus: each pound of food contained somewhat more stored energy from sunlight than humans, often with the help of animals, had to expend in growing it. That meager margin was what sustained life. Today, from farm to plate, depending on the degree to which it has been processed, a typical food item may embody input energy between four and several hundred times its food energy. This energy deficit can only be maintained because of the availability of cheap fossil fuels, a temporary gift from the Earth's geologic past.

While the application of fossil energy to farming has raised productivity, income to farmers has not kept pace. For consumers, food is cheap; but farmers often find themselves spending more to produce a crop than they can sell it for. As a result, many farmers have given up their way of life and sought urban employment. In industrialized countries, the proportion of the population that farms full-time fell precipitously during the 20th century. In 1880, 70.5 percent of the population of the United States were rural; by 1910, the rural population had already declined to 53.7 percent. In the US today, there are so few full-time farmers that census forms for the year 2000 included no such category in their list of occupations.

Mechanization favors large-scale farming operations. In 1900, the average size of a farm in Iowa was 150 acres; in 2000, it was well over twice that figure. However, the proportion of food produced by family farmers on a few hundred acres is itself dwindling; the trend is toward production by agribusiness corporations that farm thousands, even tens or hundreds of thousands of acres. In addition, a few giant multinational corporations control the production and distribution of seed, agricultural chemicals, and farm equipment, while other huge corporations control national and international crop wholesaling.

The transportation of food ever further distances has led to the globalization of food systems. Rich industrialized nations have used loans, bribes, and military force to persuade nations with indigenous populations surviving on small-scale, traditional subsistence cultivation to remove peasants from the land and grow monocrops for export. In the early part of the 20th century this practice gave rise to the phrase "banana republic," but the latter half of the century only saw the trend increase. In nation after nation, tiny subsistence plots were joined together into huge corporate-owned plantations producing coffee, tea, sugar, nuts, or tropical fruits for consumers in the US, Europe, and the increasingly prosperous countries of the Far East. Meanwhile, the ranks of the urban poor grew as peasants from the countryside flocked to shantytowns on the outskirts of places like Mexico City, Lagos, Sao Paulo, and Djakarta.

Today in North America, food travels an average of 1,300 miles from farm to plate. Consumers in Minneapolis and Toronto enjoy mangoes, papayas, and avocados year-round. In London, butter from New Zealand is cheaper than butter from Devon.

The production of meat and the harvesting of fish have likewise resulted in more energy consumption over the course of recent decades. A carnivorous diet is inherently more energy-intensive than a vegetarian diet; as growing populations in the Americas and Asia have adopted a more meat-centered fast-food diet, energy inputs per average food calorie have increased. Motorized fishing boats are much more effective at harvesting fish from the sea than their 19th-century sailing equivalents, though they are far less energy-efficient. But their very effectiveness poses a problem in that nearly all marine fisheries are now in decline as a result of overfishing.

The ecological effects of fossil fuel-based food production have been catastrophic, particularly with respect to agriculture. Farmers now tend to treat soil as an inert medium with which to prop up plants while force-feeding them chemical nutrients. As a result, the complex ecology of the living soil is being destroyed, leading to increased wind and water erosion. For every bushel of corn produced in Iowa, three bushels of topsoil are lost forever. Meanwhile, agricultural chemicals pollute lakes, rivers, and streams, contributing to soaring extinction rates among mammals, birds, fish, and amphibians.

There are signs that limits to productivity increases from industrial agriculture are already well within sight. Global per-capita food production has been falling for the past several years. Grain surpluses in the exporting countries (Canada, the US, Argentina, and the European Union) relative to global demand have disappeared, and farmers are finding it increasingly difficult to maintain production rates of a range of crops due to the salinization of irrigated croplands, erosion, the loss of pollinator species, evolved chemical resistance among pests, and global warming. For each of the past several years, world grain production has failed to meet demand, and grain in storage is being drawn down at a rate such that stocks will be completely depleted within two to five years.

Prospects for increasing food production above the global level of demand are dim, largely due to continued population growth. In his 1995 book *Who Will Feed China?: Wake-up Call for a Small Planet*, Lester Brown documents how and why China will need to import more and more grain in the decades ahead in order to feed its expanding population. Brown notes that "[a]lthough the projections ... show China importing vast amounts, movements of grain on this scale are never likely to materialize simply because they, along with climbing import needs from other countries, will overwhelm the export capacity of the small handful of countries with an exportable surplus."[2]

Add to this already grim picture the specter of oil depletion. It is not difficult to imagine the likely agricultural consequences of dramatic price hikes for the gasoline or diesel fuel used to run farm machinery or to transport food long distances, or for nitrogen fertilizers, pesticides, and herbicides made from oil and natural gas. The agricultural miracle of the 20th century may become the agricultural apocalypse of the 21st.

Expanding agricultural production, based on cheap energy resources, enabled the feeding of a global population that grew from 1.7 billion to over 6 billion in a single century. Cheap energy will soon be a thing of the past. How many people will post-industrial agriculture be able to support? This is an extremely important question, but one that is difficult to answer. A safe estimate would be this: *as many people as were supported before agriculture was industrialized* — that is, the population at the beginning of the 20th century, or somewhat fewer than 2 billion people.

There are those who argue that this figure is too low because new seed varieties and cultivation techniques developed during the past century should enable far more productivity per acre than farmers of the year 1900 were able to achieve.

This optimistic vision of the future of agriculture is currently being put forward by two camps with diametrically opposed sets of recommendations. One camp, consisting of the organic and ecological agriculture movements, recommends eliminating chemical inputs, shortening the distance between producer and consumer, and reducing or eliminating monocropping in order to support biodiversity. A recent report by Greenpeace International entitled *The Real Green Revolution: Organic and Agroecological Farming in the South* notes that in "this research we have found many examples where the adoption of [organic and ecological agriculture] has led to significantly increased yields."[3]

The other camp, led by the agricultural biotechnology industry, has proposed an entirely different solution: the genetic engineering of new crop varieties that can outproduce old ones, grow in salty soil, or yield more nourishment than traditional varieties while requiring fewer chemical inputs. According to

Hendrik Verfaille, President and CEO of Monsanto, the foremost corporate producer of gene-spliced agricultural seeds, this "technology increases ... crop yields, in some cases dramatically so. It is a technology that has been adopted by farmers faster than any other agricultural technology."[4]

Optimists in both camps assume that energy conservation and alternative energy sources will cushion the impact of fossil-fuel depletion on agriculture.

But one could argue just as cogently that the figure of two billion as a long-term supportable human population is too high. Throughout the 20th century, croplands were degraded, traditional locally adapted seed varieties were lost, and farming skills were forgotten as the number of farmers as a percentage of the population — especially in industrialized countries — waned dramatically. These trends imply that, without fossil fuels, a smooth reversion to levels of productivity seen in the year 1900 may actually be unrealistically optimistic.

Organic or ecological agriculture can be even more productive in some situations than industrial agriculture, but local success stories cannot make up for the fact that the total amount of nitrogen available to crops globally has been vastly increased by the Haber-Bosch ammonia synthesis process, which is currently dependent on fossil fuels. Ammonia synthesis could be accomplished with hydrogen, which could in turn be produced with hydroelectic hydrolysis; but the infrastructure for such production is currently almost nonexistent. It will be extremely difficult to replace all or even a substantial fraction of the added available nitrogen from ammonia via organic sources (manures and legumes). John Jeavons, of the organization Ecology Action in Willits, California, has spent the past quarter century researching methods for growing a human diet on the minimum amount of land using no fossil-fuel inputs; he has concluded that survival is possible on as little as 2,800 square feet, enabling a theoretical maximum sustainable global carrying capacity of 7.5 billion humans. However, Jeavons' "biointensive" mini-farming method assumes the composting of all plant wastes and human wastes — including human bodies *post mortem* — and provides a strictly vegan diet with no oils and no plant materials devoted to the making of fuels for cooking or heating. A more realistic post-fossil fuel carrying capacity would be substantially below the current population level.[5]

As for the genetic engineering of food crops: the technology is risky and likely to have serious unintended environmental or health consequences that could more than wipe out whatever short-term benefits it may offer. Moreover, it will not substantially reduce dependence on fossil fuels.[6]

If we simply permit the optimistic and the pessimistic arguments to cancel one another out, at the end of the day we are still left with something like two billion as an educated guess for planet Earth's sustainable, long-term, post-petroleum carrying capacity for humans. This poses a serious problem, since there are currently nearly six-and-a-half billion of us, and our numbers are still growing. If this carrying-capacity estimate is close to being accurate, then the difference between it and the current population size represents the number by which human numbers will likely be reduced between now and the time when oil and natural gas run out. If that reduction does not take place through voluntary programs of birth control, then it will probably come about as a result of famines, plagues, and wars — the traditional means by which human populations have been culled when they temporarily surpassed the carrying capacity of their environments.

Heating and Cooling

Compared to food production, heating and cooling may seem far less consequential — matters merely of comfort. However, in many places — particularly the northern regions of North America, Europe, and Asia — a source of heat can mean the difference between life and death.

Currently in the US, according to the EIA, residential energy use accounts for 21 percent of the total national energy consumption. Of this, 51 percent is consumed for space heating, 19 percent for water heating, and 4 percent for air conditioning. The rest powers lights and appliances, including refrigerators.

Modern urban life offers a context in which it is easy to take heating and cooling for granted. Fuels and electrical power are piped or wired into houses and offices sight unseen and do their work silently and predictably at the turn of a knob or the flick of a switch. Many office buildings have windows that cannot be opened, and few homes are designed for maximum energy efficiency. Temporary winter interruptions in fuel supplies often lead to deaths; and during the summer, elderly people who lack access to air conditioning are vulnerable to extreme heat. In an average year in the US, 770 people die from extreme cold and 380 from extreme heat; combined, these figures exceed the average combined death tolls from hurricanes, floods, tornadoes, and lightning.[7] Serious and continuing fuel shortages would probably lead to a substantial increase in mortality from both temperature extremes.

Natural gas is widely used in industrialized countries for cooking and for heating hot water; diminishing supplies will obviously result in higher costs for these services. Only a relatively small proportion of the total amount of natural gas used goes toward cooking; however, since this is an essential function, a protracted interruption in supplies could have a major impact on people's daily lives.

Energy in the form of electricity is the primary power source for the refrigeration of food. As electricity becomes more expensive due to shortages of natural gas and the decline in net energy from coal, refrigeration will become more costly. Without refrigeration, supermarkets will be unable to keep frozen foods, and produce will remain fresh for much shorter time periods. The food systems of cities will need to adjust to these changes.

In areas of the world where wood, other plant materials, and dried animal wastes are used as fuel for space heating and cooking, air pollution and deforestation are already serious problems. Thus, given present population densities, the substitution elsewhere of such traditional fuels for oil and natural gas will pose serious environmental and health hazards.

The Environment

The energy transition of the coming century will affect human society directly, but it will also likely have important indirect effects on the natural environment.

Some impacts — such as deforestation from increased firewood harvesting — are relatively easy to predict. As fossil fuels become scarce, it will become increasingly difficult to protect trees in old-growth forest preserves, and perhaps even those along the sides of city streets.

Other environmental effects of oil and natural gas depletion are less predictable. It is tempting to speculate about the impact on global warming, but no firm conclusions are possible. At first thought, it might seem that fossil-fuel depletion would actually improve the situation. With ever fewer gallons of gasoline and diesel fuel being burned in the engines of cars and trucks, less carbon dioxide will be released into the atmosphere to contribute to the greenhouse effect. Perhaps petroleum depletion could accomplish what the Kyoto protocols on greenhouse gas emissions have only begun to do.

However, it is important to remember that when global oil production peaks, half of nature's original endowment of crude will still be in the ground waiting to be pumped and burned. Extraction rates will gradually taper off but will not suddenly plummet. If efforts are made to increase coal usage in order to offset energy shortages from oil and natural gas, greenhouse gas emissions might remain close to current levels or even rise. Thus, unless a coordinated, intelligent program is put in place for a transition to non-fossil energy sources as well as for a rapid and drastic curtailment of total energy usage, the net effect of oil and natural gas depletion on the problem of global warming is not likely to be significantly positive over the next few decades.

The situation is similar with regard to the problem of chemical pollution: a decline in the extraction of fossil fuels might seem to hold the promise of reducing environmental harms from synthetic chemicals. With less plastic being produced and fewer agricultural and industrial chemicals being used, the load of toxins on the environment should decrease. However, many pollution-monitoring, -control, and -reduction systems currently in place — including trash pick-up and recycling services — also require energy. Thus, even if the production of new chemicals declines, over the short run there may be heightened problems associated with the containment of existing pollution sources.

The reduced availability of oil and natural gas will likely provoke both electrical energy producers and politicians to call for a reduction of pollution controls on coal plants and for the building of new nuclear plants. But these strategies will entail serious environmental costs. Increased reliance on coal, and any relaxation on emissions controls, will result in more air pollution and more acid rain. And increased reliance on nuclear power will only exacerbate the unsolved problem of radioactive waste disposal.

As the global food system struggles to come to terms with the decline in available net energy for agriculture, transportation, and food storage, people who have the capacity to fish or to hunt wild animals will be motivated to do so at increasing rates. But given mounting energy and financial constraints, conservation agencies will find it difficult to control overfishing and the over-hunting of edible land animals. Endangered species will have fewer protections available and extinction rates will likely climb.

The environmental impacts of changing patterns in agriculture are difficult to predict, given that the direction of those changes is uncertain. If efforts are made to localize food production and to voluntarily reduce chemical and energy inputs via organic/ecoagricultural methods, then the current detrimental environmental impacts of agriculture could be reduced markedly. However, if the managers of global food systems opt for agricultural biotechnology and attempt to sustain inputs, negative environmental effects from food production are likely to continue and, in the worst case — a biotech "frankenfood" disaster — could be catastrophic.

In sum: it is possible to imagine scenarios in which the decline in fossil-fuel extraction and consumption could, on balance, be relatively good for the environment, or very bad indeed. It will all depend on how governments and other institutions choose to respond.

Public Health

International, national, and local systems of public health, which protect the human population against communicable diseases and parasites, are also vulnerable to declines in the availability of cheap energy. Water and sewage treatment, medical research, and the production and distribution of antibiotics and vaccines all require power. In the next few decades, unless the percentage of total available money and energy devoted to public health increases, more- as well as less-industrialized societies will face at worst severe epidemics and at best increased disease-related death rates.

Today, infectious diseases already cause approximately 37 percent of all deaths worldwide. Waterborne infections account for 80 percent of all infectious diseases globally, and 90 percent of all infectious diseases occur in the less-consuming countries. Each year, a lack of sanitary conditions contributes to approximately 2 billion human infections causing diarrhea, from which 4 million infants and children die. Even in industrialized nations, waterborne diseases pose a significant health hazard: in the US they account for 940,000 infections and approximately 900 deaths each year.[8]

Approximately 1.2 billion people in less-consuming nations lack clean, safe water. Of India's 3,119 towns and cities, just 209 have partial treatment facilities and only 8 have full wastewater treatment plants; 114 cities dump untreated sewage and partially cremated bodies directly into the sacred Ganges River.

Many diseases that can easily and cheaply be treated or prevented still pose problems in many areas of the world. New strains of *E. coli* are spreading in parts of Africa and Asia where humans are crowded and where water and food contamination is rampant. At least 300 million acute cases of malaria occur globally each year, resulting in more than a million deaths, most of them in Sub-Saharan Africa. Moreover, tuberculosis is on the rise in many nations due to crowding and drug resistance. Currently, an estimated 1.7 billion people worldwide are infected with TB, with approximately 95 percent of deaths occurring in less-consuming countries. In 1990, the annual number of new TB infections was 7.5 million; by 2000, the number had reached 10 million.[9]

Human plague — which is assumed to have been the disease that decimated European societies throughout the medieval period — continues to break out periodically. The plague parasite, *Yersinia pestis*, is transmitted by human contact with rodents. In the 1980s, the average of the annually reported cases in the world was 1,350; in the 1990s, the average annual number rose to 2,500. Nearly 60 percent of the reported cases occurred in Africa.

Diphtheria had been under control for many years; but, following the breakup of the former Soviet Union, the disease made a startling comeback. In 1975, about 100 cases were recorded in Russia; but in 1995 alone, 51,000 new cases were reported. The World Health Organization attributes this recent explosion in diphtheria in Russia to a decline in the effectiveness of that nation's public health program.

In the decades ahead, global warming will likely contribute to the spread of infectious tropical diseases such as malaria, putting a further strain on already over-taxed public health systems.

Meanwhile, as many long-familiar diseases that were formerly in decline are making a comeback, entirely new diseases continue to arise, including hantavirus, Lyme disease, Creutzfeldt-Jakob disease, Legionnaire's disease, West Nile virus, ebola haemorrhagic fever, Venezuelan haemorrhagic fever, Brazilian haemorrhagic fever, and AIDS. The last of these poses perhaps the greatest public-health challenge in the world today.

In eastern and southern Africa, HIV infection is cutting down an alarming percentage of Africa's most energetic and productive adults aged 15 to 49. In 2001, more people on the continent succumbed to HIV than to any other cause of death, including malaria. While only 10 percent of the world's population lives in sub-Saharan Africa, the region is home to two-thirds of the world's HIV-positive people and has suffered more than 80 percent of all AIDS deaths. In Zaire and Zimbabwe, more than a quarter of the adults carry the virus. In a few districts, rates of infection approach 60 percent. If infection rates continue to grow unchecked and if mortality figures follow infection rates, AIDS will soon dwarf every catastrophe in Africa's recorded past.[10] AIDS cases are now being reported in rapidly increasing numbers in Russia and China as well.

In short, global public health systems are already taxed beyond their limits.[11] But what will be the impact of a reduced energy availability on those under-funded and over-extended systems?

It could be argued that the impact of oil depletion on the medical and health infrastructure need not be severe since many public-health problems (such as those stemming from lack of clean water) can theoretically be solved relatively cheaply. Moreover, even if the end of oil and natural gas were to mean turning back the clock of technological development to pre-industrial levels, that would not necessarily imply the loss of all intervening advances in medical science. Anesthesia, antiseptics, surgery, and transfusions save tens of thousands of lives annually and need not disappear with reduced energy availability.

Nevertheless, modern medicine taken as a whole is a highly energy-intensive enterprise. A hospital in a typical industrial city uses more energy per square foot of space than nearly any other kind of building. As the interval of cheap energy wanes, the wealthy few will likely continue to have access to modern forms of care for their various health problems, but even the richest countries will find it increasingly difficult to support the development or distribution of new vaccines or of new antibiotics to combat the rapidly emerging strains of resistant diseases.

While the severity of the public-health problems the next generation will face is impossible to estimate, worst-case scenarios are truly horrific.

In any event, the medical profession will need to adapt to an entirely new energy environment, with all that this implies in terms of changes in transportation and other forms of support infrastructure; urban authorities will need to find less energy-expensive ways to maintain water treatment and waste disposal facilities and otherwise ensure public hygiene; and national governments will need to make deliberate efforts to channel a much greater percentage of available energy and money away from other sectors (such as the military) and toward public health, if a steep increase in preventable deaths is to be averted.

Information Storage, Processing, and Transmission

Electronic information technologies — including computers, telephones, fax machines, computer printers, and internet servers — are critical to the functioning of modern industrial societies. They have come to play essential roles in the coordination and management of financial, commercial, manufacturing, medical, and military systems, so that a general failure of data and communications systems would soon imperil much of the support infrastructure of society as a whole.

The daily operation of information technologies is not, to any appreciable degree, directly dependent on oil. Thus the peak in petroleum extraction will not have an immediate impact on information storage, processing, and transmission. However, the construction, maintenance, and distribution of the components of information systems do depend, to a much larger extent, on oil-fed transportation and on the fabrication of plastics. Thus, over the long term, oil scarcity will make it more difficult to maintain or expand current information systems.

However, the daily operation of the information infrastructure of industrial societies *is* directly dependent on regional electrical grids. These grids are complex, costly to maintain, and highly vulnerable to interruptions in the supply of basic energy resources (coal, uranium, and natural gas) used to generate electricity. Moreover, demand for electricity continues to increase, fed partly by continued population growth. As the net energy available to industrial societies wanes, resources devoted to the electrical grids will become relatively more expensive. At a certain point, demand for electricity will begin consistently to exceed supply. From then on, the electrical power grids may become threatened. Periodic brownouts and blackouts may become common. These may in turn interrupt the critical functions of information systems. In that case, the only way to maintain the grids would be to increase the price of electricity sufficiently to discourage nonessential uses, and this would have a significant impact on the economy. Within years of the first widespread blackouts it may become impossible to maintain the grids at their present scope, and efforts may be made to reduce the size of grids and to cannibalize components that can no longer routinely be replaced. Eventually it might no longer be possible to maintain the electrical grids in any form.

If and when that point is reached, *unless an alternative renewables-based electrical infrastructure is already substantially in place,* the information infrastructure of industrial societies will collapse and virtually all electronically coded data will become permanently irretrievable.

National Politics and Social Movements

The implementation of the most intelligent strategies for dealing with the petroleum extraction peak — such as diverting remaining energy resources toward conservation and transition efforts — will require political will. But politicians are seldom inclined to deal with problems proactively, and will be unlikely to act decisively until crisis has arrived full-blown. Moreover, due to their perennial need for large campaign contributions, politicians (particularly in the US) are much more likely to respond to the advice of wealthy corporate leaders than to that of scientists or citizens; and corporate leaders in turn customarily take their cues primarily from economists — who tend to discount even the possibility of resource shortages in their confidence that the all-knowing market will magically provide substitutes for whatever commodities become scarce.

Thus the current shape of the political landscape will affect how we deal with — or fail to deal with — the energy transition. And at the same time, the energy transition will change the political landscape in profound, structural ways.

Politics is, at least in part, the social contest for control over resources. The current political scene has resulted from long-term resource rivalries between relatively empowered and disempowered social groups. As energy supplies dwindle, those rivalries will be greatly exacerbated.

According to the political theory of the Right, each individual is morally entitled to gain control over as large a share of the total resource base as he or she can possibly obtain, using legal means. Traditional ways of expanding one's resource share include capturing energy from other humans by hiring them for wage labor (from which surplus value is extracted in the form of profits) and by investing in energy-leveraging productive enterprises that depend directly or indirectly on energy resources extracted from the Earth. The government, according to this theory, has little or no responsibility to maintain equity in resource distribution or to provide a safety net for the disadvantaged. The Right thus gains part of its legitimacy from its appeal to the individual's desire for freedom — the freedom, that is, to control a disproportionate share of resources. Most great cultural achievements, according to rightists, have been initiated not by the masses but by extraordinary individuals. Thus it is by giving rein to the individual quest for accomplishment and gain that society as a whole is bettered.

Another cornerstone of rightist politics is the pursuit of security through state investments in ever-expanding police and military powers. Such powers are needed, after all, to protect the concentrations of wealth that result from the project of seeking to control resources. The Right aims its appeal primarily at those with disproportionate wealth, and secondarily to members of the lower classes who envy the wealthy or who can be persuaded that the highest aims of the state are law, order, and security.

According to the political theory of the Left, people are morally obliged to share resources and society as a whole is better off when inequalities of wealth are minimized. Here, government has a responsibility to help equalize access to resources and to provide at least a minimum of needed resources for all citizens. Theorists at the far-left end of the political spectrum hold that all exploitation of humans by other humans — including the system of wage labor — is morally repugnant. The Left typically seeks to appeal to members of the lower classes and to idealistic intellectuals. While the Communist-bloc nations of the 20th century offered a counter example, the Left has historically been somewhat less preoccupied than the Right with police and military security;

and where it calls for criminal sanctions or military actions, these are often against individuals, corporations, or nations whose efforts to control resources appear egregiously unfair.

Democracy — the social means whereby citizens collectively and consciously control the conditions of their lives — is often regarded as an artifact of Greek civilization or the Enlightenment; but from a larger historical and anthropological perspective it can be seen as an attempt on the part of people living in modern complex societies to regain some of the autonomy and egalitarianism that characterized life in the hunter-gatherer bands of our distant ancestors. Democracy is a reaction against the concentrations of power that arose in early agricultural states and that burdened our more recent ancestors with kingship and serfdom. Because it implies that everyone should be able to participate in decisions regarding the allocation of resources, democracy is an inherently leftist ideal. This remains true despite the profoundly undemocratic nature of the Communist-bloc societies of the 20th century. Unquestionably, the most innovative thinking regarding democratic processes has come from the far-left of the political spectrum, which is occupied by anarchists of various stripes.

Both leftist and rightist ideologies contain an element of unreality or even denial concerning population and resource issues. Most rightists preach that *all* who wish for success and who work hard can potentially be wealthy if each individual is freed to compete in the market, unimpeded by government regulation. Most leftists promise that, if wealth is shared and decisions are made cooperatively, there will be plenty for everyone — with no exceptions in the face of population pressure or resource depletion. A few rightists acknowledge resource limits but argue that, since existence is a Darwinian struggle anyway, it is the fit (the wealthy) who should survive through economic competition while the unfit (the poor) are culled by starvation. A few leftists acknowledge limits but believe that, if humanity is made aware of them and empowered to deal with distribution issues democratically, people will decide to undertake a process of voluntary collective self-restriction that will enable everyone to thrive within those limits. Typically, when either leftist or rightist regimes actually encounter resource limits, some aspect of ideology (democracy on the one hand, the free market on the other) is sacrificed, at least to some extent.

The contest between the Left and the Right is probably an inevitable dynamic within every civilization, but it has developed into its current form only since the late 18th century — that is, since the start of the industrial interval.

Throughout the energy upswing, the political contest ebbed and flowed through periods of colonialism, anti-colonialism, populism, socialism, communism, fascism, the Cold War, and corporate globalization, with all sides competing for rights to an *expanding* base of available resources. As petroleum extraction peaks and energy resources become more scarce, the entire political landscape will shift as both the Right and the Left try to come to terms with the new reality.

Particularly in wealthy nations, the Right will no doubt seek to exploit people's heightened competitiveness and their felt need for security and authority during a time of flux. The general populace, seeing the world coming apart at the hinges, fearing a breakdown of law and order, and wanting to know whom to blame for mounting economic ills, will likely rally to strong leaders who offer scapegoats and who promise to maintain order by whatever means necessary.

Meanwhile the Left will appeal to people's moral indignation at the wealthy and powerful, who maintain extremely unequal shares of resources even as vast numbers of humans lose access to basic necessities. Many citizens will also be outraged at corporate and political leaders for their failure to anticipate the obviously inevitable energy transition and for their failure to inform and warn the public.

The net-energy decline will bring challenges to all social groups, both the empowered and the disempowered. The wealthy will find it difficult to maintain social control and to justify extreme inequality. Leftists will be seeking an equal share of a shrinking pie — which will lead to the feeling of having metaphorical goal posts continually moved backward as one approaches them. With victory always receding toward the horizon, the rank and file may find it hard to maintain their morale. Moreover, as rising expectations confront dwindling realities, leftists in wealthier countries (such as the US) may be branded as traitors to the cause of maintaining their nation's unequal control of global resources.

Since it is easier to contemplate sharing when there is plenty to go around than when what little one has is disappearing, population pressure and resource scarcity will likely place ever-greater stress on the already battered democratic ideals of industrial societies.

If the Right gains the upper hand, the result will probably be the undermining of civil liberties; the scapegoating of leftists, minorities, and foreigners; and the expansion of military and police powers. Democracy will become a ritualized sham at best. If the Left gains the upper hand, the result might be a kind of modern peasant revolt, in which the wealthy will be demonized and punished.

However, neither political response will necessarily do much to solve the underlying problem of energy-resource depletion.

Because they have no solution, politicians on both sides will probably go to absurd lengths to obscure or mystify the real causes of the changes engulfing society. The public will likely not hear or read much about peaks in the extraction rates of oil or natural gas. They will see prices for basic commodities increase sharply (in inflation- or deflation-adjusted terms), but the ensuing economic turmoil will be held to be the fault of this or that social, political, ethnic, national, or religious group, rather than being identified as the unavoidable result of industrialism itself. The Left will blame selfish rich people and corporations; the Right will blame foreigners, "terrorists," and leftists.

Many people already sense that the traditional political categories of Left and Right no longer hold the solutions for today's unique social and environmental problems. Sociologist Paul Ray has argued, on the basis of extensive polling data, that a sizable portion of European and American populations consist of "cultural creatives" who defy both leftist and rightist stereotypes.[12] These are people who typically espouse ecology and feminism while questioning globalization and the power of big business. It is conceivable that this constituency, if united and mobilized, could press for sensible energy policies.

The signal political development of the past decade has been the emergence of the global-justice movement advocating "globalization from below." That movement, to which many cultural creatives are drawn, demands the democratization of all social institutions and the limitation of the power of corporations to exploit workers in less-consuming countries; it also envisions a borderless world in which people can move without restriction. As the project of corporate globalization collapses for lack of energy resources, the anti-globalizationists will see their warnings about the consequences of undermining local economies fully vindicated. However, corporate leaders may blame the global-justice movement for having helped cause the collapse of the global economy. And with dwindling resources motivating growing hordes to migrate en masse seeking necessities for survival, the ideal of a borderless world may seem less attractive to the settled segments of the populace.

In order for the movement to meet the challenges of the post-petroleum era, it must discard all socioeconomic analysis rooted in the 19th century (e.g., classical Marxism and some strains of anarchism), which assumes industrial growth based on increasing energy-resource availability. The analysis needed today must take into account ecological principles, energy-resource constraints, population pressure, and the historical dynamics of complex societies — including the infrastructural reasons for their growth and collapse. This analysis has already begun within some quarters of the environmental movement, but even there it is neither complete nor widely disseminated.[13]

The movement's intellectual leaders will be tempted to seize on the new energy constraints as evidence of mismanagement on the part of the government-corporate authorities (which, of course, is the case), but then to withhold the crucial information that the new energy regime (entailing shortages, economic chaos, and general suffering) is by now a historical inevitability. They will find it difficult to resist the incentive to offer the public promises of plenty, if only the reins of political power are shifted. The alternative — telling the public the awful truth that the era of cheap energy and industrial growth is over — may be politically unpalatable, but in the long run it is the only morally defensible course of action: *the sooner the general public understands the situation industrial societies are in, the less suffering will occur as we make the inevitable but painful transition to a new energy regime.*

Over the past few decades and in the US particularly, rightist forces have so successfully inoculated the public with corporate-funded propaganda that an open debate between Left and Right over how to respond to the emerging crisis may never take place.[14] Instead, the Right may simply reign triumphant.

But if growing public dissatisfaction arising from the shrinking of the resource base is denied coherent expression through a leftist alternative, it will seek some other outlet. It could, for example, be expressed through increased intergenerational conflict. Even if not explicitly told that this is the case, young people will likely intuitively understand that, within the lifetime of the baby-boomer generation, nearly half of the total petroleum reserves of the planet were used up. Everywhere they will see evidence of the extravagant party their elders have thrown, while for themselves there will be only dregs left over. With ever fewer economic opportunities available, they may feel an unspeakable resentment toward older people who have frittered away the world's endowment of natural resources, leaving almost nothing for their children and grandchildren. If rightist forces are powerful enough to prevent this rage from being channeled into an organized leftist movement, young people may vent their anger through random acts of sabotage, which will only provoke and justify increased repression.

Over the long term, however, the prospects for maintaining the coherence of large nation states like the US, regardless of the philosophy governing their political apparatus, appear dim. Lacking an industrial infrastructure of production, transportation, communication, and control, large nations may eventually devolve into regional enclaves — which, depending on the local circumstances, could have political structures that are either democratic or authoritarian, depending on local circumstances.

The Geopolitics of Energy-Resource Competition

Just as political rivalries *within* nations will be exacerbated by the energy transition, so those *between* nations will be heated to the boiling point.

Resource conflicts are nothing new. Pre-state societies often fought over agricultural land, fishing or hunting grounds, horses, cattle, waterways, and other resources.[15] As we saw in Chapter 2, most of the wars of the 20th century were also fought over resources — in some cases, oil. But those wars took place during a period of expanding resource extraction; the coming decades of heightened competition over fading energy resource supplies will likely see even more frequent and deadly conflicts.

Though it is an empire in steep decline, the US — as the world's largest energy consumer, the center of the global industrial empire, and the holder of the most powerful store of weaponry in world history — will nevertheless play a pivotal role in shaping the geopolitics of at least the first decades of the new century. It is therefore probably best to begin an exploration of international relations during the net-energy decline with a survey of current US geopolitical strategy, especially as it relates to energy resources.

For the past few decades, the US has pursued a dual policy in the Middle East, the most oil-rich region of the planet. On the one hand, it has supported repressive Arab regimes in order to maintain access to petroleum reserves. America persuaded its Arab oil-state clients to denominate their production in US dollars. By thus being required to pay for most of their oil imports in dollars, importing countries around the world have contributed a subtle tithe to American banks and the US economy with every barrel of crude purchased. Arab rulers take a share of the "petrodollar" profits from the oil extracted from their countries and channel much of what they receive toward investments in the West and toward the purchase of US weapons. In exchange, they have been promised US protection against their own people, who would naturally prefer to benefit more directly from the immense energy wealth with which nature has endowed their lands.

On the other hand, the US has supported Israel unquestioningly and with vast amounts of money and weaponry. Especially after its impressive military victory over the Arab states in 1967, Israel came to be seen as a foil to Arab nationalism. The Nixon Doctrine defined Israel's role as that of the "local cop on the beat," serving US military and intelligence interests in the region. This role became still more important following the Iranian Revolution in 1979, which denied the US its other main base in the Middle East. Israel also serves to deflect Arab resentment away from the United States: even though the US extracts considerable wealth from Arab countries, until recently Arab anger

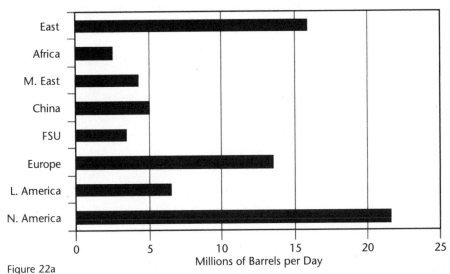

Figure 22a

Oil consumption by region, in millions of barrels per day. (Source: C. J. Campbell)

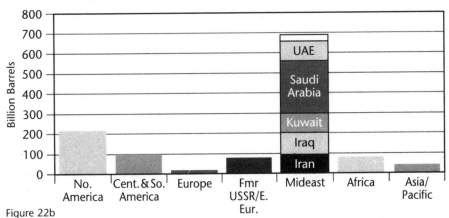

Figure 22b

Oil reserves by nation Jan. 1, 2003. North American reserves include Canadian tar sands; Mideast figures are stated reserves, which are the subject of controversy. (Source: Oil & Gas Journal, 12/23/02)

tended to be borne primarily by Israel, with the US depicting itself as a friend to the Arabs. American-backed Arab leaders likewise used Israel as a foil to divert the anger of the so-called "Arab street" away from their regimes' corruptness and toward the neighboring Jewish state.

Both long-standing US Middle-East policies are fraying. The two have had a history of mutual tension in any case; the most prominent instance of conflict between them was the (Arab-only) OPEC oil embargo against the West in late 1973 and early 1974, which was provoked by the US support for Israel during the Arab-Israeli war of 1973 and which resulted in severe economic hardship for the American economy.

Since September 11, 2001, the disharmony between these two policies has become truly cacophonous. The position of the rulers in Saudi Arabia and most of the small Gulf States is gradually being undermined, as these regimes have come under increasing public criticism in the US since September 11. The current Bush administration is no doubt reassuring these rulers that the US will continue to guarantee their security, but they must nevertheless view the shift in American public opinion as worrisome. Their pariah status in the eyes of the American people in turn makes it more difficult for the monarchies to ignore the sentiment of their own populace — which is growing increasingly critical both of Israel's policies regarding the Palestinians, and of the US "war on terrorism." The invasion and ongoing occupation of Iraq add yet more division and uncertainty.

At the same time, the US leaders have announced to the world that they will unilaterally decide which actions around the world constitute "terrorism" and which do not. In the view of Arabs and Muslims everywhere, the US appears to have concluded that all actions by Palestinians against Israelis, whether against Israeli soldiers or against innocent civilians, constitute "terrorism." The US (and this includes most of the US media as well as the government) has also evidently adopted the attitude that no acts on the part of the Israeli government against Palestinians constitute "terrorism." Arabs view this as a double standard; for this reason as well, expressions of distrust and hatred of the US are mounting. When the Arab peoples see their own governments supporting America's self-defined "war on terrorism," their antagonism toward their US-backed rulers intensifies. The rulers, seeing this deepening antagonism, are becoming increasingly uneasy.

It is impossible to say whether any of the governments in the oil-rich Arab nations whose stability and security the US has guaranteed for nearly 60 years will collapse in the near future. However, those governments are clearly now under greater internal stress than at any time in the past few decades. The Saudi royal family appears divided as to the line of succession from the ailing King Fahd. Moreover, most of the citizens of Saudi Arabia subsist largely on state subsidies derived from oil revenues, which have been falling partly due to population expansion. This fall in payments to the young can hardly help but cause social tensions within that country. Thus, Saudi Arabia may well be headed toward turmoil, which could lead the US to intervene to seize the oil fields in the eastern part of that country. It seems at least possible that one of the purposes of the Iraq invasion was, in fact, to position new permanent military bases within easy striking distance of the Saudi fields.

Indeed, it may be that the disastrous outcome of the Iraq invasion has left the US with reduced, rather than expanded, options in the region.

While it is impossible to get inside the minds of US geopolitical strategists, statements by American officials suggest that they are at least contemplating the option of maintaining open supply lines through a variety of means tailored to the current realities in each of the world's oil-rich regions.

The Middle East: Next to the oil fields of Saudi Arabia, the world's largest petroleum reserves are those of Iraq, which were more or less withheld from the world market during the decade of sanctions , thus helping to keep global oil prices from falling too low.

Since the US-led invasion of Iraq, ongoing turmoil and outright sabotage have prevented the further development of that country's oil fields and export infrastructure to any appreciable degree. Indeed, in many months since the invasion, oil exports have lagged behind levels seen during the latter years of Saddam Hussein's regime.

To say that the Iraq occupation has not gone well is a serious understatement, and the consequences are likely to be grim. The US cannot simply leave, because to do so would create a power vacuum that could lead to political chaos throughout the region. But maintaining the present course will likely result in expanding resistance and civil war within the country, which could, in turn, lead to political chaos throughout the region. In short, it is difficult at this point to imagine a sequence of events leading to a peaceful and constructive outcome — barring some dramatic and unforeseeable change of strategy on the part of the US.

At the same time, regimes throughout the Middle East are on edge, seeking to rein in simmering anti-American sentiment among the Arab population, in order not to provoke further US overt or covert actions that would destabilize their governments.

Iran is likely to be a nexus of struggle in the near future. The US and Europe wish to deter Iran from developing nuclear weapons — which the Iranians see as essential to deterring American imperialist aggression. Meanwhile, both China and Russia are cooperating with Iran increasingly in the areas of energy and mutual defense. From a geopolitical perspective, Iran bridges the oil-rich regions of Middle East and Central Asia, lying adjacent to Iraq on the west and Afghanistan on the east. Iran is also a major oil and gas producer, and is thus crucial to the futures of importing nations. Moreover, the Iranian government has voiced interest in selling its energy resources in currencies other than the US dollar.

The Caspian Sea: Next to the Middle East, the Caspian Sea region contains perhaps the world's largest untapped reserves of both oil and natural gas (though these have probably been over-estimated). Most of these reserves are in territory that was formerly part of the Soviet Union; some are bordered by Iran. In order to be marketable, these reserves must be accessible by pipeline. American geopolitical strategists are concerned not that these resources necessarily end up in the US, but that the US government and US-based corporations be in a position to control their flow, and hence their price.

This requires keeping any new pipelines from passing through Iran, which is still an oil power and is still operating independent of US control. American officials prefer an expensive route through Turkey to the Mediterranean, and another route through Afghanistan to Pakistan. In May 2005, the $4 billion Baku-Ceyhan-Tblisi pipeline — which begins in Azerbaijan and passes through Georgia and Turkey but bypasses Russia and Iran — opened, delivering one million barrels of crude per day to mostly European markets. Shortly after the recent US military action in Afghanistan, agreements were signed for the Afghan pipeline. However, despite backing from the World Bank and local governments, that project may be years from completion; current diplomatic efforts in that regard are essentially directed toward consolidating power in the region. This goal is also being sought through the same strategies used in the Middle East — by buying off corrupt regimes with promises of security and with shipments of arms for potential use against unruly civilian populations. Meanwhile the US has built 19 new military bases in the Caspian region, which appear to be permanent fixtures of the "war on terror."

Russia would prefer to see much of the Caspian's oil and gas flow north rather than south. Moreover, Russia has its own considerable oil and gas reserves and, even though its petroleum production peaked in 1987, it is in fact probably far better situated in that regard than the US over the short term. Currently it is exporting fossil fuels to Europe and the industrial nations of Asia. Throughout the 1990s, US leaders sought to use loans and debt to turn Russia into a dependent client state, and partly succeeded in that effort. However, Russian leaders are aware of the ace they hold in terms of their remnant military and industrial infrastructure, and their relatively abundant fossil-fuel reserves. Thus while the US and Russia remain overtly on friendly terms, the possibility of renewed geopolitical rivalry lurks close to the surface.

Under Vladimir Putin, Russia is seeking to regain some of the geopolitical prowess of the old Soviet Union. The privatization of industries and resources that occurred under Yeltsin has declined — as symbolized by the quashing of efforts by executives to sell Yukos (one of the largest Russian oil companies), to Western firms.

Russia's greatest advantage may lie simply in its geography: it is a vast country that covers much of the landmass of Eurasia.

If the US is to remain the world's superpower, it must dominate Eurasia, the site of two-thirds of the world's energy resources. This will be difficult to accomplish from thousands of miles away. America's oil imports must arrive by tanker, and this is an inherently vulnerable supply chain. The maintenance of imperial outposts likewise implies vulnerable supply chains stretching across oceans. In contrast, the countries of Eurasia can rely on pipelines, and on alliances based on geographic proximity. From a geostrategic point of view, an alliance between Russia, Europe, and perhaps China would be America's ultimate nightmare.

But this is exactly what is emerging, and the US has only itself to blame. The unilateralism of the Bush administration has predictably provoked collaborative activity on the part of countries that feel frozen out of decisions that affect their interests.

Barring an escalating confrontation over Iran, this geopolitical rivalry is not likely to erupt into a shooting war any time soon, but economic warfare seems nearly inevitable at this point. And here again, the US is extremely vulnerable, as concerted action by only a few nations could easily result in the severe undermining of the value of the US dollar.

South America: Venezuela is America's third largest oil supplier, and a prominent member of the Organization of Petroleum Exporting Countries (OPEC).

Soon after his election in 1998, Venezuela's president, Hugo Chavez, passed a spate of new laws that, among other things, increased the government's share of revenue from oil exports. Chavez also reformed Venezuela's constitution, through a constitutional assembly and a referendum, making it one of the most progressive constitutions in the world.

Given this record, the US-backed April 11, 2002 coup attempt against Chavez seemed wholly predictable. However, a successful counter-coup three days later reinstated Chavez, proving him to be a resourceful and resilient politician.[16]

If Chavez sticks to his quasi-leftist principles, the US will likely search for other ways to reassert control over Venezuela's oil wealth. Further coup attempts are highly likely. Meanwhile, however, China is bidding for access to Venezuela's oil.

Meanwhile, in Colombia, the US has increased military aid to the regime of Alvaro Uribe Velez — ostensibly to help the Colombian army root out cocaine growers and smugglers. However, it is clear to nearly all international observers that another, perhaps more pressing goal is to secure US corporate interests — including oil fields, pipelines, and coal mines — from rebels in the country's 40-year-old civil war.[17]

China: The world's most populous nation possesses indigenous energy resources, but not on a scale large enough to fuel its accelerating process of industrialization. Continued reliance on domestic coal supplies has economic advantages, but it will entail environmental devastation and will be incapable of powering the development of China's transportation infrastructure. With its burgeoning appetite for energy, China is capable of dramatically changing the global supply/demand picture for oil and natural gas. Until recently, the US provided the marginal demand in crude oil. But now China is building refineries at a rapid rate, even as its consumption of crude far outpaces its indigenous production. China is using Dickensian sweat shops and near-slave labor in order to grow its economy; but its leaders know that, in order for its efforts at industrialization to succeed, human labor must increasingly be tied to fuel-fed machinery.

China has recently surpassed Japan to become the world's second foremost oil importer (the US is still first in line, importing twice as much as China and Japan combined). Increasingly, China and the US are competing for long-term oil export contracts in Central Asia, the Middle East, Africa, and even Canada and South America.

China's economic influence is expanding quickly throughout Asia — including the contested Caspian Sea region — bringing it inevitably into conflict with

US strategic interests there. Here as elsewhere, American strategists would prefer to avoid direct confrontation, as China's increasing share of the global economy and its massive production of export goods for the US market ensure that any open conflict would inevitably harm both sides. Nevertheless, since China is capable of absorbing a quickly growing share of the available global oil exports, economic and possibly military conflict with the US is likely sooner or later.

Economic warfare between the two nations would damage both severely. The US has been able to run up massive deficits in recent years partly because of China's willingness to purchase American government debt in the form of Treasury Bills. China could thus help precipitate a collapse of the US dollar merely by dumping its investments on the international market. However, this would hurt China as well, since that country is dependent on food imports from the US, which could be halted if competition turns ugly.

China also has strategic energy-resource interests in the South China Sea that overlap with those of nations other than the US. The area — bordered on the north by China, on the east by the Philippines, on the south by Indonesia and Malaysia, and on the west by Vietnam — is believed to possess significant undersea resources of gas and oil (though exploration efforts to date have been disappointing). All of the nations in the region have conflicting claims on those resources. As policy analyst Michael Klare has pointed out in his book *Resource Wars: The New Landscape of Global Conflict*,

> growing demand for energy in Asia will affect the South China Sea in two significant ways. First, the states that border on the area will undoubtedly seek to maximize their access to its undersea resources in order to diminish their reliance on imports. Second, several other East Asian countries, including Japan and South Korea, are vitally dependent on energy supplies located elsewhere, almost all of which must travel by ship through the South China Sea. Those states will naturally seek to prevent any threat to the continued flow of resources. Together, these factors have made the South China Sea the fulcrum of energy competition in the Asia-Pacific region.[18]

In recent years, China has seized several islands from Vietnam and established military outposts on them; meanwhile, most of the nations in the region have embarked on an arms race to protect shipping lanes and defend resource claims.

Britain: Only a few years ago British Prime Minister Tony Blair was hailing the new information economy as a replacement for the old oil economy. Then came the oil price spike of 2000, which wreaked temporary havoc on London's financial markets. Perhaps Blair has since come to appreciate the significance of the fact that the rate of his nation's share of the North Sea oil and gas extraction appears to have peaked in 1999–2000. North Sea oil gave the UK a tremendous economic boost during the past three decades; but as of 2005 Britain has ceased to be an oil exporter and will need to import increasing amounts of petroleum in coming years in order to maintain its economy. British coal production is also in steep decline. Blair may have decided that his nation's economic survival hinges on future access to the resources of the Middle East and that the best way to ensure that access is through maintaining a close military and political alliance with the US. Though his positions on issues in this regard are often unpopular among his constituents, Blair is forced by circumstances to provide the US with aid and cover in its otherwise unilateralist pursuit of global resource dominance.[19]

The Balkans: This is not a resource-rich region, but one essential to the transfer of energy resources from Central Asia to Europe. It is also the site of Camp Bondsteel, the largest "from-scratch" foreign US military base constructed since the Vietnam War. Located in the Yugoslav province of Kosovo on farmland seized by US forces in 1999, Camp Bondsteel lies close to the US-sponsored Trans-Balkan oil pipeline, which is now under construction. Brown and Root Services, a Houston-based contractor that is part of the Halliburton Corporation, the world's largest supplier of products and services to the oil industry, provides all of the support services to Camp Bondsteel — including water, electricity, spare parts, meals, laundry, and firefighting services.[20]

While it would no doubt be an oversimplification to say that US military action in the Balkans in the 1990s was motivated solely by energy-resource considerations, it might be just as wrong to assume that such considerations played only a minor role.

Regional rivalries and long-term strategy: Even without competition for energy resources, the world is full of conflict and animosity. For the most part, it is in the United States' interest to prevent open confrontation between regional rivals, such as India and Pakistan, Israel and Syria, and North and South Korea. However, resource competition will only worsen existing enmities.

As the petroleum production peak approaches, the US will likely make efforts to take more direct control of energy resources in Saudi Arabia, Iran, the Caspian Sea, Africa and South America — efforts that may incite other nations to form alliances to curb US ambitions.

Within only a few years, OPEC countries will have control over virtually all of the exportable surplus oil in the world (with the exception of Russia's petroleum, the production of which may reach a second peak in 2010, following an initial peak that precipitated the collapse of the USSR). The US — whose global hegemony has seemed so complete for the past dozen years — will suffer an increasing decline in global influence, which no amount of saber rattling or bombing of "terrorist" countries will be able to reverse. Awash in debt, dependent on imports, mired in corruption, its military increasingly overextended, the US is well into its imperial twilight years.

Meanwhile, whichever nations seek to keep their resources out of the global market will be demonized. This has already occurred in the cases of Iran, Iraq, and Libya — which sought to retain too large a share of their resource profits to benefit their own regimes and hence attained pariah status in the eyes of the US government. Essentially they were seeking to do something similar to what the American colonists did in throwing off British rule over two centuries earlier. Like the American colonists, they wanted to control their own natural resources and the profits accruing from them. Many readers will object to such an analogy between American colonial patriots and modern-day Libyan or Iraqi leaders on the grounds that the latter are, or were, autocrats guilty of human-rights abuses that justified their condemnation by the international community. But we must recall that America's founders were themselves engaged in slavery and genocide and that many US client states — including Turkey, Israel, Indonesia, and Saudi Arabia — have also been guilty of serious abuses.[21]

In the future, secure access to resources will depend not only on the direct control of oil fields and pipelines but also on successful competition with other bidders for available supplies. Eventually, the US will need to curtail European and Japanese access to resources wherever possible. Again, every effort will be made to avoid direct confrontation because in open conflict all sides will lose. Even the closest trading partners of the US — Canada and Mexico, which are currently major energy-resource suppliers — will become competitors for their own resources when depletion reaches a point where those nations find it hard to maintain exports to their energy-hungry neighbor and still provide for the needs of their own people.

Civil wars will be likely to erupt in the less-industrialized nations that have abundant, valuable, and accessible resources, such as oil, natural gas, and diamonds, rather than in those that are resource-poor. This conclusion is based on a correlation study by Indra de Soysa of the University of Bonn of the value of natural resources in 139 countries and the frequency of civil wars since 1990.[22] The finding runs counter to the long-held assumption that internecine warfare is most likely to occur in resource-poor countries. Often rival groups within nonindustrial countries use wealth from the sale of resources — or from leases to foreign corporations to exploit resources — in order to finance armed struggles. Pity the nations with resources remaining.

The least industrialized of the world's nations will face extraordinary challenges in the decades ahead, but may also enjoy certain advantages. Industrialized nations will seek to choke off the flow of energy supplies to resource-poor economies, most likely by yanking their debt chains and enforcing still more structural-adjustment policies. However, less-industrialized nations are able to squeeze much more productivity out of energy resources than are the energy-saturated economies of the industrialized nations. Less-industrialized nations are therefore potentially able to bid prices higher, or to absorb higher energy costs much faster, than the industrialized nations. This is only one of many wild cards in the longer-term game that will be played out as the world's energy resources slowly dribble away.

Taking It All In

This is probably a good point to stop and take a breath. The picture drawn in this chapter is a profoundly disturbing one. It depicts a century of impending famine, disease, economic collapse, despotism, and resource wars. The reader may be wondering: Is the author deliberately exaggerating the perils ahead in order to make a point? Or is he simply a gloomy and depressed individual projecting his neuroses onto the world?

Nothing in this chapter was written deliberately to depress or alarm. The future projections under each heading above represent not possible though improbable disasters — like an asteroid striking the Earth tomorrow — but the likely outcomes of present trends. And I hasten to point out that, while my personal life has held its share of frustrations and disappointments, I am reasonably cheerful and optimistic by nature. However, as anyone would, I find this picture of the future to be deeply disturbing. Everyone I have met who understands population and resource issues comes to essentially the same conclusions and has to deal with the same emotional responses — which typically run the gamut from shock, denial, despair, and rage to eventual acceptance — and a determination to do whatever is possible to help avert the worst of the likely impacts.

Richard Duncan of the Institute on Energy and Man reached essentially these same conclusions when he began to correlate world energy use and population data in terms of overshoot and collapse. His resulting "Olduvai theory" predicts that the life of industrial civilization will be a "horridly short" pulse lasting roughly 100 years (from 1930 to 2030), with its high point corresponding to the peak of global per-capita energy use — which occurred in 1979.[23] He named his theory after the Olduvai Gorge in Tanzania, which is associated in the public mind with human origins and the Stone-Age way of life. Duncan believes that humanity will return to an essentially Stone-Age existence after the end of fossil fuels and industrialism (I don't agree with Duncan that this is the inevitable outcome of the energy transition, since many civilizations existed before fossil fuels came into use). "Industrial civilization doesn't evolve," Duncan writes. "Rather, it rapidly consumes the necessary physical prerequisites for its own existence. It's short-term, unsustainable." After developing his theory for over a decade, Duncan now thinks that "electricity is the quintessence of industrial civilization" and that it will be the failure of the power grids, rather than the peaking of global oil production, that will trigger the end of industrialism.

What was Duncan's emotional response to his own theory? He writes:

> Back in 1989 I became deeply depressed when I concluded that our greatest scientific achievements will soon be forgotten and our most cherished monuments will crumble to dust. But more so, I knew that my children would feel the pressure, and will likely suffer. That really hurt. In time, however, my perspective changed. Now I just treat the Olduvai theory like any other scientific theory. Nothing personal. Each year, I gather the data ... and watch the theory unfold.

But why should anyone pay attention to the gloomy prognostications of population/resource analysts in the first place, when there are so many cheerier images of the future available from economists, politicians, and religious leaders? Realistically, human nature being what it is, I assume that the vast majority of people will continue to prefer happy illusions to the stark truth, no matter how compelling the arguments in this or any other book on energy resources. Still, the fact remains: as long as we trade on false hopes, we only dig deeper the hole we're already in.

We tend to be victims of what Ernest Partridge of the University of California at Riverside calls "perilous optimism." As Partridge puts it,

When, Exactly, Does the Party End?

It is probably simplistic to equate the coming peak in petroleum production with the end of industrialism. There are at least six major linked events that could be considered markers of the end of the historic interval of cheap energy, and two of them have already occurred:

1. **The peak in global per-capita energy production.** According to White's Law, "culture evolves as the amount of energy harnessed per capita per year is increased, or as the efficiency of the instrumental means of putting energy to work is increased."[25] During the period from 1945 to 1973, world energy production per capita grew at 3.24 percent per year. From 1973 to 1979, growth slowed to .64 percent per year. From 1979 to 2000, energy production per capita declined at an average rate of .33 percent per year. However, growth of energy demand in China and India in 2003 and 2004 resulted in a spike in global per-capita energy consumption that reached the 1979 level.[26]

2. **The peak in global net-energy availability.** Throughout the past couple of decades, more total energy has continued to be produced each year, on average, from all sources combined; but the amount of energy spent in obtaining energy has increased at a faster pace. This is especially true for oil, coal, and natural gas, for which net yields are falling precipitously: it requires more drilling effort to obtain a given quantity of gas or oil now than it did only a few years ago, and more mining effort to obtain the same amount of coal. The peak in the total net energy available annually worldwide has almost certainly already passed, but it is unclear exactly when: complex calculations are involved and no official agency has bothered to undertake them. A good guess would be that the net-energy peak occurred between 1985 and 1995.

3. **The peak in global oil extraction.** As discussed in Chapter 3, this peak will probably be reached between 2006 and 2010. The exact year is uncertain, and the event may be masked or altered by economic factors. We will know only in retrospect exactly when the peak occurred.

4. **The global peak in gross energy production from all sources.** This is likely to coincide closely with the global oil-extraction peak. ☞

[h]uman beings thrive on hope. Without some sense that our individual deliberate effort brings us closer to a fulfillment of our personal goals, we simply cannot function from one day to the next. And yet, hope often betrays us, as it blinds us to clear and evident danger and leads us to courses of action and inaction that will eventually result in the loss of our property, our livelihood, our liberty, and even our very lives.[24]

5. **The energy-led collapse of the global economy.** Even if an economic collapse occurs first for other reasons (as fallout from the collapse of the US dollar, the bursting of the American stock-market bubble, expanding war in the Middle East, or the implosion of more scandal-ridden American corporations), energy constraints will eventually hit the global financial system. Energy scarcity will cause a recession of a new kind — one from which anything other than a temporary, partial recovery will be impossible. We humans may, if we are intelligent and deliberate, create a different kind of economy in the future, building steady-state, low-energy, sustainable societies characterized by high artistic, spiritual, and intellectual achievements. But the industrial-growth global economy that we are familiar with will be gone forever. The timing of this event will again depend upon that of the global petroleum production peak.

6. **The collapse of the electricity grids.** This collapse may occur at somewhat different times, and at different rates, in different nations and regions, depending on the robustness of the grids themselves, on the resource basis with which electricity is generated (coal, nuclear power, hydro, wind, etc.), and on the continued local availability of particular fuels. For example, the decline in natural gas production in North America may hasten grid failure in this part of the world. But everywhere, except in regions where electrical power is already supplied mostly from renewable sources (and such places are rare), the grids are extremely vulnerable; given the time and the investment levels needed to switch to renewable sources of electricity on a large scale, even if extraordinary efforts are undertaken now the electrical generation and distribution systems on which industrial societies depend may ultimately be unsustainable.[27] If and when they come down for good, it's lights out. The party will truly be over. ■

If optimists see the glass as half full and believe that things are good and getting better, they may conclude that there is little need to be concerned about the future and hence fail to take action. On the other hand, when pessimists see the glass as half empty and believe the world is going downhill and getting worse every day, they may conclude that there is nothing that can be done and also fail to take action. It is the realists who, seeing that society faces dire and increasing threats, recognize that there is much that can be done to mitigate the worst of the likely impacts and take informed action to make the best of the situation.

That is my essential purpose in this chapter — not to depress but to help readers who are willing to do so to face reality squarely and to take informed action, so that as many as possible of the dire impacts discussed here can be prevented or mitigated.

Those who live in industrialized countries happen to have been born into the most complex societies in history, ones that have reached the stage where — as Joseph Tainter would put it — the returns on their ongoing investments in greater complexity are quickly diminishing. We have arrived at a point where global societal collapse — meaning a reversion to a lower level of complexity — is likely, and perhaps certain, over the next few decades. Once humanity has passed through the coming period of shedding complexity, it is entirely possible that our descendants will attain a much less-consuming, fulfilling way of life. But the process of getting from here to there is likely to be horrendously difficult, and the desirability of the outcome will depend to a very high degree on actions taken now.

6

Managing the Collapse:
Strategies and Recommendations

We are all addicts of fossil fuels in a state of denial, about to face cold turkey. And like so many addicts about to face cold turkey, our leaders are now committing violent crimes to get what little is left of what we're hooked on.

— Kurt Vonnegut

We must face the prospect of changing our basic ways of living. This change will either be made on our own initiative in a planned way, or forced on us with chaos and suffering by the inexorable laws of nature.

— Jimmy Carter (1976)

To avoid deprivation resulting from the exhaustion of nonrenewable resources, humanity must employ conservation and renewable resource substitutes sufficient to match depletion.

— Ron Swenson (2001)

We can't conserve our way to energy independence, nor can we conserve our way to having enough energy available. So we've got to do both.

— George W. Bush (2001)

If we accept the notion that the global industrial system will probably collapse in one way or another within the next few decades, several questions follow. Some inevitably center on personal survival and the welfare of family and friends. Others are more generally humanitarian in spirit: How can we

225

minimize human suffering as the party winds down? How can we preserve as much as possible of nature and culture? Further, how can we find a way down the Hubbert curve that offers incentives and satisfactions so that the human spirit will still have worthy goals (other than continued economic growth and material affluence) toward which to strive?

If collapse cannot be avoided altogether, the best alternative is clearly a *managed collapse*, in which society would undertake a deliberate, systematic process of simplifying its structures and reducing its reliance on nonrenewable energy sources. (Again: I am using the term *collapse* here in the technical sense in which Tainter employs it, namely to refer to any substantial reduction in social complexity, and not necessarily to the complete, sudden, chaotic disintegration of all institutions.)

There is already an extensive literature of recommendations along these lines — although some of it seriously understates the political and economic challenges inherent in the project of deliberately shrinking the material throughput of a social system designed on the assumption that resource availability will continually grow.

One of the better recent texts in this regard is *Beyond the Limits: Confronting Global Collapse, Envisioning a Sustainable Future,* by Donella Meadows, Dennis Meadows, and Jørgen Randers (1992). In it, the authors present the updated results of their computer model World3 which, in the early 1970s, modeled future outcomes from trends in population and resource use, producing projections of industrial collapse in the mid-21st century (this initial work was reported in 1972 in the best-selling book *The Limits to Growth*). When Meadows et al. refined the program and fed in new data twenty years later, they again found that "the model system, and by implication the 'real world' system, has a strong tendency to overshoot and collapse. In fact, in the thousands of model runs we have tried over the years, overshoot and collapse has been by far the most frequent outcome."[1]

In spite of this, the authors believe that a "sustainable society is still technically and economically possible. It could be much more desirable than a society that tries to solve its problems by constant expansion."[2] They offer recommendations based on their computer modeling that could enable industrial societies to avoid destructive collapse if programs of resource conservation, population stabilization, equitable goods distribution, and emissions reduction were adopted immediately (that is, by the mid-1990s). The authors write that it is:

impossible for anyone now to describe the world that could evolve from a sustainability revolution as it would have been for an English coal miner of 1750 to imagine a Toyota assembly line. The most anyone can say is that, like the other great revolutions, a sustainability revolution could lead to enormous gains and losses. It too could change the face of the land and the foundations of human self-definitions, institutions, and cultures. Like the other revolutions, it will take centuries to develop fully — though we believe it is already underway and that its next steps need to be taken with urgency .... [3]

Another helpful set of recommendations is offered by environmental systems analyst Hartmut Bossel in his book *Earth at a Crossroads: Paths to a Sustainable Future* (1998). There Bossel contrasts the current "competitive model" of unsustainable development with a "partnership model" of sustainable development. Continued competition, according to Bossel, will lead to resource wars, a reduction of resources available per capita, and a polarization of rich and poor. Cooperation, on the other hand, holds the promise of networking the unique capabilities and skills of people of diverse backgrounds to achieve a synergy, such that the whole is greater than the sum of its parts. Thus, in terms of human welfare, more could be achieved with less matter and energy flowing through the social system. Bossel writes:

> In discussing our future, it is important that we understand the full implications of "sustainability" .... A sustainable society will have to allow development without physical growth (of material and energy flows and population). Its population must eventually remain below a certain limit that is probably less than today's global population. The per capita use of energy and materials must be less than what it is now in the industrialized countries of the North. All energy must be renewable, all materials recyclable. These limited throughputs of resources must support a system that maintains an unlimited potential for non-material cultural, social, and individual growth. [4]

Essentially similar arguments are made by Howard T. Odum and Elisabeth C. Odum in their book *A Prosperous Way Down: Principles and Policies* (2001). They note that

> [p]recedents from ecological systems suggest that the global society can turn down and descend prosperously, reducing assets, population,

and unessential baggage while staying in balance with its environmental life-support system. By retaining information that is most
important, a leaner society can reorganize itself and continue making progress .... The reason for the descent is that the available
resources on Earth are decreasing .... That the way down can be
prosperous is the exciting viewpoint whose time has come. Descent
is a new frontier to approach with zeal .... If everyone understands
the necessity of the whole society adapting to less, then society can
pull together with a common mission to select what is essential ....
The alternative is a world of selfish battles for whatever resources
remain.[5]

Virtually all of the authors who have contributed to the literature on sustainability tell us that, in order for a transition to a lower-complexity and
lower-throughput society to occur without a chaotic collapse, humanity will
have to take a systemic approach to resource management and population
reduction.

In this final chapter I intend to sketch the general outlines of the social, economic, political, and individual-lifestyle changes that are needed in order to
minimize the consequences of energy-resource depletion and to build the
foundations of a society capable of enduring for many generations into the
future. I will save two important questions — *Is it too late?* and, *Are these recommendations realistic?* — until the end of the chapter.

You, Your Home, and Your Family

There is much that you, as an individual, can do to prepare for the energy transition. Below are suggestions grouped into eight categories; but as you take
your first steps on the path toward a sustainable lifestyle, you will find that
these strategies naturally blend into each other. You will also find new friends
who are on the same path and who can offer encouragement and suggestions.
Many thousands of people find satisfaction in making these sorts of efforts for
their own sake — not just as a strategy for survival during an anticipated social
or economic crisis. Over the past decade or so, my wife and I have employed
most of these strategies in our home and with our community of friends; and
my colleagues and students and I explore them in some detail in our yearlong
program on Culture, Ecology, and Sustainable Community at New College of
California.

Energy usage. Begin by assessing your current energy usage, then decide which areas of usage are essential and which are nonessential. Gradually and deliberately reduce your nonessential usage. This is a process that may continue over some time and may require considerable experimentation and ingenuity. Examine your utility bills carefully and begin using them as a feedback mechanism to tell you how you are doing in your conservation efforts.

You can improve the energy efficiency of your home relatively easily by replacing incandescent lights with compact fluorescents; by more thoroughly insulating walls and roof; by replacing single-pane glazing with high-e double-pane windows; and by choosing energy-thrifty appliances. Direct most of your effort toward the area where your energy usage is greatest. For most people, this will be home heating.

Alternative energies. After you have pared your energy usage to the bare minimum, consider equipping your home with a renewable energy source. Photovoltaic systems are expensive now, but when electricity prices begin to soar you might be glad you invested in one.

Wind power may be feasible for you if you live in a rural setting. Small wind turbines generate power more efficiently than do PV panels, but they require tall towers and can make an unpleasant noise.

If you rent your house or apartment, altering the building itself may seem unfeasible. You may instead wish to examine your housing options: might it make sense to move to a place where it would be easier to pursue radical energy efficiency and energy independence?

Your home. If you are thinking of building a new home or remodeling your existing one, consider using ecological design principles and natural or recycled materials. Straw-bale, rammed-earth, and cob construction can be used to build houses that stay warm in the winter and cool in the summer with little or no energy usage. Many counties now routinely grant building permits for these kinds of alternative structures, which have proven themselves over time to be durable and efficient.

Building one's own structure is an extraordinarily empowering experience. If you lack construction skills, take a workshop on basic carpentry and find a builder who is familiar with natural building, who is willing to teach you, and who will allow you to do as much of the work as you can.

If you live in a rural or semi-rural area, a composting toilet might be a good alternative to a conventional septic system, in that it would allow you to use human wastes as fertilizer for trees and shrubs. As discussed by Joe Jenkins in his *Humanure Handbook,* there are even simpler and more direct methods for composting human waste, though local ordinances typically prohibit them.[6]

Finances. Reduce your debt. Whatever interest you are paying on loans — especially credit-card interest — is nonproductive and a drain on your personal energy budget. Further, don't buy what you don't absolutely need. Forget about your "patriotic duty" to the consumer economy and to the maintenance of the national financial system. Your primary duty is to a higher cause: personal and planetary survival.

Exiting the consumer treadmill is psychologically as well as financially freeing. The "voluntary simplicity" movement has been growing internationally for the past two decades, and local networks and support groups exist in many areas.

Appropriate technology. Begin replacing some of the Class D tools in your life with Class A, B, and C tools (see Chapter 1, pp. 25, 26). A well-made hand tool — a hoe, garden spade, saw, chisel, or plane — is a joy to use; employing it properly requires skill, but offers considerable satisfaction.

These days it is often more time-consuming and expensive to repair and reuse manufactured objects than simply to throw them away and replace them. But as energy resources become more scarce and valuable, having basic maintenance and repair skills could mean the difference between continual frustration and lack on the one hand and sufficiency and satisfaction on the other. Many junior colleges offer classes to the public on small-motor repair. Knowing how some of the simple devices we depend on actually operate tends to raise one's level of self-confidence, even in the absence of energy shortages. Begin to assemble a small library of books and articles on home repair and maintenance and begin to try fixing simple things on your own, seeking advice whenever necessary.

Health care. Perhaps the most important appropriate technologies are those for health maintenance. Learn about healing herbs and basic medical procedures that can save lives in the temporary absence of doctors and hospitals. Start a medicinal herb garden in your back yard or window box and assemble a natural home medicine chest consisting of dried herbs and herbal tinctures, as well as books on natural first-aid remedies.

Food. Grow as much of your own food as you can. Doing so successfully will require practice and experimentation: gardening is both an art and a science. If you live in an apartment, explore window-box or hydroponic gardening.

Unless you have a very large city lot or some acreage, considerable gardening experience, and a fair amount of time on your hands, it will be unrealistic to expect to grow all of the food you will need to sustain yourself and your family. However, you can make it your goal to grow more of your diet each season by managing your garden more carefully, and by planting a wide variety of vegetables that can be harvested more or less continuously. A greenhouse or cold frame can help extend your growing season year-round.

Saving seed is a time-honored traditional craft that contributes both to self-reliance and to the maintenance of the genetic commons. Buy open-pollinated, non-hybrid varieties of vegetable seeds, and in your garden set aside some space where a few plants from each variety can complete their life cycles, yielding seeds for next year. Seek out neighbors who are avid gardeners and whose families have lived in your area for several generations: they may have heirloom seed varieties, well adapted to your local soil and climate, that they would be happy to share with you in return for some of your own more unusual seeds or produce.

Look for alternatives to chemical fertilizers, pesticides, and herbicides. Some nurseries specialize in supplies for the organic gardener.

In order to keep from quickly depleting your soil, you will need to build and renew it each year. You can make your own compost from lawn clippings, leaves, soil, manure, kitchen scraps, and crops grown especially for the compost pile. A worm box — which turns kitchen scraps into rich black humus — can be employed even in a small apartment.

If you have the space, keeping a few chickens can serve several purposes at once: chickens can produce both food (eggs and, if you wish, meat) and nitrogen-rich fertilizer while periodically ridding your garden of snails, slugs, and invasive insects.

Food self-reliance entails devoting some thought and effort toward preservation and storage. Drying is the easiest means of preservation, and it requires no energy source other than the Sun. Canning takes more planning, work, and energy, but enables you to put up larger quantities.

It is easy to construct a solar oven that will cook food even on cold winter days, as long as the Sun is out. A meal consisting entirely of food that you have grown and that you have cooked slowly in a solar oven is truly a banquet of self-sufficiency.

Transportation. For most Americans, the automobile is a key personal link to the oil-based energy regime of industrialism. A new car rolls off the assembly line each second, and the global fleet uses twice as much energy from oil as humans obtain from the food they eat.

Resources for Home and Family

Energy Usage

The Energy Saving House: A Practical Guide to Saving Energy and Saving Money in the Home, by Thierry Salomon and Stephane Bedel (New Society, 2002).

The Home Energy Diet: How to Save Money by Making your Home Energy Smart, by Paul Scheckel (New Society, 2005)

Alternative Energies

The Solar Living Sourcebook, John Schaeffer, ed. This catalog for solar and wind technologies as well as for composting toilets and energy-efficient appliances is produced by Real Goods, and is updated periodically. <www.realgoods.com>

Solar Today magazine 2400 Central Ave., Suite G-1, Boulder, CO 80301, USA. <www.ases.org/soltoday/>

Home Power magazine, P.O. Box 520, Ashland OR, 97520, USA. <www.homepower.com>

Home

The Art of Natural Building: Design, Construction, Resources, Joseph E. Kennedy, Michael G. Smith, and Catherine Waneck, eds. (New Society, 2002).

Earthbag Building: The Tools, Tricks and Techniques, by Donald Kiffmeyer and Kaki Hunter (New Society, 2004).

Finances

Your Money or Your Life: Transforming Your Relationship with Money and Achieving Financial Independence, by Joe Dominguez and Vicki Robin (Penguin, 1999).

Stepping Lightly: Simplicity for People and the Planet, by Mark A. Burch (New Society, 2000).

Radical Simplicity: Small Footprints on a Finite Earth, by Jim Merkel (New Society, 2003).

The Simple Living Network

Appropriate Technologies

When Technology Fails: A Manual for Self-Reliance and Planetary Survival, by Matthew Stein (Clear Light, 2000).

Microhydro: Clean Power from Water, by Scott Davis (New Society, 2003). ☞

Health Care

Where There Is No Doctor: A Village Health Care Handbook, by David Werner (Hesperian Foundation, 1992).

Natural First Aid: Herbal Treatments for Ailments and Injuries, Emergency Preparedness, Wilderness Safety, by Brigitte Mars (Storey, 1999).

Herbs for the Home Medicine Chest, by Rosemary Gladstar (Storey, 1999).

Food

The Sustainable Vegetable Garden: A Backyard Guide to Healthy Soil and Higher Yields, by John Jeavons and Carol Cox (Ten Speed Press, 1999).

The Solar Food Dryer: How to Make Your Own Low-Cost, High-Performance Sun-Powered Food Dryer, by Eben Fodor (New Society, 2005).

Permaculture is a holistic approach to sustainable homestead systems design, integrating food, housing, water, and energy. Look for local classes.

Ecology Action seed catalog, from Bountiful Gardens, 18001 Shafer Ranch Road, Willits, CA 95490, USA

<www.growbiointensive.org/biointensive/Gardening.html>

Harmony Farm Supply specializes in supplies for the organic gardener. Catalog: 3244 Hwy. 116 North, Sebastopol, CA 95472, USA

Plans for a home-build solar oven are available at

<http://solarcooking.org/plans.htm>

Transportation

Divorce Your Car!: Ending the Love Affair with the Automobile, by Katie Alvord (New Society, 2000).

From the Fryer to the Fuel Tank: The Complete Guide to Using Vegetable Oil as an Alternative Fuel, by Joshua Tickell and Kaia Roman (Greenteach, 2000). ∎

Consider the possibility of living car-free. When Katie Alvord, author of *Divorce Your Car!*, went on a national book tour, she traveled by bus, train, and folding bicycle just to prove that, even in our auto-dependent society, it can be done. As a first step, go car-free one day a month, then one day per week. Drive only when necessary and walk, use mass transit, or car-pool whenever possible. Vancouver and San Francisco now have car co-ops: if you live in either of these cities, join one; if you don't, start one. If you must have a personal vehicle, give some thought to the kind of car you drive. When it comes time to replace your metal monster, consider the alternatives: buying an older used car will entail a smaller energy cost than buying a new one, and buying a gas-thrifty model may save both fuel and money. Electric and hybrid cars are now available, and it is possible to operate a diesel vehicle on biodiesel fuel — or, after a small alteration to the fuel system, to run it on recycled vegetable oil that can be obtained at little or no cost.

Your Community

The strategy of individualist survivalism will likely offer only temporary and uncertain refuge during the energy downslope. True individual and family security will come only with community solidarity and interdependence. If you live in a community that is weathering the energy downslope well, your personal chances of surviving and prospering will be greatly enhanced, regardless of the degree of your personal efforts at stockpiling tools or growing food.

During the energy upslope, most traditional communities became atomized as families were torn from rural subsistence farms and villages and swept up in the competitive anonymity of industrial cities. Security during the coming energy transition will require finding ways to reverse that trend.

The first steps can be taken immediately. Get to know your neighbors. Look for people in your community who share a similar interest in voluntary simplicity and self-reliance, and begin to form friendships and habits of mutual aid.

Your community may be a village, a neighborhood, or a city. Regardless of its size, you will find opportunities for building alliances. There are inevitable challenges inherent in this project: your community no doubt includes people whose interests conflict with your own. Rather than identifying "wedge" issues and highlighting disagreements, find areas of common interest that are related to the goal of community sustainability and security regarding universal human needs: food, water, and energy.

The Post Carbon Institute (www.postcarbon.org) has prepared materials to assist with creating community responses to peak oil (see sidebar p. 240). Consider starting a Post Carbon "outpost" in your town or city. Begin by hosting community meetings on the subject, then organize action committees.

Food. Few communities bother to examine the security and sustainability of the food system on which they rely. Berkeley, California, is one of the first cities to deliberately undertake such an assessment. In 2001, the Berkeley City Council passed the Berkeley Food and Nutrition Policy, which provides the community with a clear framework for the next decade to help guide its creation of a system that, in the words of its mission statement, is "based on sustainable regional agriculture," that "fosters the local economy," and that "assures all people of Berkeley have access to healthy, affordable and culturally appropriate food from non-emergency sources." Berkeley's schools have adopted a policy of serving organic food to students and identified the goal of having a garden in each school.

Community gardens provide good food and build community at the same time by transforming empty lots into green, living spaces. Members of the community share in both the maintenance and rewards of the garden. Today there are an estimated 10,000 community gardens in US cities alone.

A particularly innovative community gardening project was begun in Sonoma County, California, in 1999 by a group of idealistic college students. Calling their loosely-knit organization Planting Earth Activation (PEA), the leaderless collective offers to dig and plant gardens for anyone free of charge; all they ask in return is a share of open-pollinated seed saved from those gardens, with which they can plant still more gardens. Garden plantings occur in a party atmosphere, with music and food to accompany the hard work. PEA chapters have recently sprung up in neighboring counties as well.

Community-supported agriculture (CSA) is a fast-growing movement that operates on the premise that the consumer contracts directly with a farmer. The CSA model of local food systems began 30 years ago in Japan, where a group of women concerned about the increase in food imports and the corresponding decrease in the farming population initiated a direct purchasing relationship between their group and local farmers. The concept traveled first to Europe and then to the US. As of January 2000, there were over 1,000 CSA farms across the US and Canada. CSA members cover a farm's yearly operating budget (for seeds, fertilizer, water, equipment maintenance, labor, etc.) by purchasing a share of the season's harvest, thus directly assuming a portion of the farm's costs and risks. In return, the farm provides fresh produce throughout the growing season.

Water. Unless you have a well or live next to a stream or lake, access to water is a more of a community issue for you than an individual one. And since water inevitably flows — both above and below the ground surface — all water issues are ultimately community issues.

Because water treatment plants and pumping stations use energy, communities will need to conserve water and find new ways to distribute water and prevent water pollution as energy resources become more precious.

Some communities have already made some efforts along these lines. After suffering through years of drought, Santa Barbara County, California, instituted a Water Efficiency Program, which offers information on home and landscape water conservation as well as educational materials on water conservation. And a grassroots group in Atlanta, Georgia, has initiated Project Harambee: they distribute free ultra-low-flush toilets, low-flow shower heads, and energy conservation information to low-income households in an effort to reduce water and energy consumption. There are similar programs in dozens of other towns and cities across North America.

Watershed protection groups and Water Watch programs also exist in regions throughout the US, monitoring rivers and streams and identifying sources of pollution. Members collect data, which they share with county and state agencies, and educate the public through literature, classes, and tours.

Natural waste-water treatment facilities, which rely on the purifying characteristics of marsh plants, are operating successfully in Germany, Switzerland, and the Netherlands; in the US, a natural waste-water treatment facility began operations in Arcata, California, in 1986. Arcata's project uses a marsh system to provide both secondary treatment for the city's waste water and wildlife habitat.

Despite initial steps like these, few communities are prepared to meet energy-based challenges to their ability to supply clean water to citizens. The development of alternative low-energy water delivery and treatment systems in large and small communities everywhere will require creativity and cooperative effort. As communities begin to prioritize their energy budgets, they will need to devote whatever power they obtain from renewable sources, such as photovoltaics and wind, first toward water, as a foundation for their collective survival and sustainability.

Local economy. Corporate globalization has hit local economies hard. In town after town, local businesses have succumbed to "big box" chains like

Wal-Mart, which buy in huge quantities and often sell mostly imported items made by low-paid workers. Once a local economy has been destroyed by dependence on the "big box," the chain frequently pulls out, forcing members of the community to drive tens of miles to the nearest larger town for basic consumer needs. Several communities have successfully resisted Wal-Mart; their campaigns have relied on group initiative, hard work, and hired consultants.

Resistance to chains must be accompanied by efforts to promote and sustain local enterprises: locally owned bookstores, restaurants, grocery stores, clothing stores, and product manufacturers. "Buy local" advertising campaigns can help keep regional economic infrastructures robust, enabling them better to face both the immediate threat of competition from national chains and the approaching challenge of the energy transition.

Since national currencies are based on debt, their use subtly but inevitably saps wealth from local communities. Every dollar loaned into existence requires the payment of interest, some of which (even if the loan is issued by a local credit union) goes to a nationally chartered banking cartel. Debt-based money thus systematically transfers wealth from the poor to the rich. In addition, national currencies are subject to inflation, deflation, and collapse as well as to manipulations and panics beyond the community's control. One solution is the promotion of local barter systems; another is the creation of a local currency. Both are legal (if operated within certain guidelines) and have long histories of success.

Public power. In many cities, electricity and natural gas are delivered by publicly owned not-for-profit power utilities, as opposed to investor-owned utility companies. There are currently roughly 2,000 municipal power districts in the US, which together deliver electricity to 15 percent of the population.

Public power enables every citizen to be a utility owner, with a direct say in policies that affect not only rates and service but choices as to energy sources as well. Citizens can, for example, decide to phase out nuclear plants and replace them with wind or solar plants — as happened in the case of the Sacramento, California, Municipal Utility District (SMUD).

Starting a publicly owned power electric utility takes considerable time, money, and effort. Nevertheless, currently several large cities and many small towns in all parts of the US are considering establishing their own utilities in order to save money and provide citizens with more control. More than 40 public power utilities have been formed in the last two decades.

The necessary steps in forming a public power utility vary from state to state, but typically include authorizing a feasibility study; analyzing pertinent local, state, and federal laws; obtaining financing; informing and involving the public; holding an election to let voters decide on the merit of the proposal; and issuing bonds to buy present facilities or finance the construction of new ones.

Citizens in many states have formed energy co-ops, which buy electricity in quantity and sell to members at a discount. Energy co-ops are private, independent electric utility businesses owned by the consumers they serve. Distribution cooperatives deliver electricity to the consumer, while generation-and-transmission cooperatives (G&Ts) generate and transmit electricity to distribution co-ops. Currently in the US, 866 distribution and 64 G&T cooperatives serve 35 million people in 46 states. In distribution co-ops, members can decide what sources to buy from: coal, nuclear power, or renewables; G&T co-op members can choose what kinds of plants to invest in. Initial investment money can often be obtained from a local credit union.

Community design. Towns and cities are continually changing, and most communities have some process in place to plan their future direction of change. When citizens become involved in the urban planning process and bring with them the values of sustainability and conservation, important strides can be made toward successfully weathering the energy transition.

During the last century, most towns and cities grew around the priorities of the automobile. Today, it is essential that communities be redesigned around people. Public transportation, walking, and bicycling must be emphasized — and car traffic discouraged. A first step is the creation of car-free zones in mid-town areas. Often such zones are a boon to local businesses; and as anyone knows who has visited old European cities like Venice or Siena, a car-free town or town center offers far more space for cultural expression than is possible in a car-dominated city. Neighborhoods can be made more pedestrian-friendly with speed bumps, snaky curves in roads, and prominent crosswalks. And towns can systematically reduce automobile traffic with carpool-only lanes. Meanwhile, funds can be diverted away from road building and toward the provision of light-rail service.

Cities and towns can also be encouraged to build more bike paths and bike lanes. Some cities are already far ahead of others in this regard. Portland, Oregon, for example, has a fleet of refurbished old bicycles available for free use within the city, showers and changing rooms for bike commuters, buses and trains that accept bikes on board, and a 140-mile bike trail encircling the city.

Usually, urban design priorities like these are articulated and promoted by citizens' advocacy groups. Concerned citizens have established sustainability groups in several cities and counties (including Sustainable Seattle, Sustainable San Francisco, and Sustainable Sonoma County); they perform sustainability studies using the Ecological Footprint indicator[7] and develop and advocate plans to improve the community's environmental health, to stabilize its local economy, and to achieve a more equitable distribution of resources. Sustainability groups examine issues of air quality, food, hazardous materials, waste, water, biodiversity, parks, and open spaces; their working groups include representatives from city agencies, local businesses, and academia.

The "new urbanism" movement, discussed in the documentary "The End of Suburbia,"[8] advocates making cities more pedestrian friendly by building up multi-use urban cores and neighborhoods and discouraging strip-mall corridors. A more radical approach is advocated by architect Richard Register, author of *Ecocities: Building Cities in Balance with Nature* (Berkeley Hills, 2001), who understands the challenge of the imminent oil production peak and envisions redesigning cities to virtually exclude the personal automobile.

Local governance. One way to make change within your community is to get involved in local politics. Politics is about power and decision-making, and it entails conflict and hard work. When the stakes are high, the political process almost inevitably becomes subject to corruption, and cynicism and burnout usually follow. However, in smaller communities, local government is still relatively open to citizens' input and participation.

Involvement in local politics opens many possibilities for moving your community in the direction of sustainability. Find out what local issues are. Go to meetings of your city council or board of supervisors. Identify people and groups with concerns similar to yours and work with them to form research and action committees.

It is possible to influence officeholders by writing letters, actively participating at public meetings, or writing opinion pieces for the local newspaper. It is also possible to run for office — even though it takes a certain kind of personality to want to do so. This requires knocking on doors, but it also presents an opportunity to educate the public.

Intentional communities. For many centuries, idealists have sought to create a better world by building model communities in which alternative ways of living can be experimented with and demonstrated. There are thousands of

Community Resources

Food

Information on community-supported agriculture (CSA):

and <www.umass.edu/umext/csa/>.

Growing Communities Curriculum: Community Building and Organizational Development through Community Gardening, by Jeanette Abi-Nader, Kendall Dunnigan, and Kristen Markley (The American Community Gardening Association, 2001).

Information on the Berkeley Food Policy Council:

and <www.pmac.net/organic_lunch.html>

Information on Planting Earth Activation (PEA):

<www.pacificsites.net/~mec/NEWSL/ISS37/37.07PEA.html>

Water

Water-efficiency success stories can be found at:

<www.sustainable.doe.gov/efficiency/water/wesstoc.shtml>

Water Watch programs: < www.waterwatchonline.org/>

Information on the Arcata natural waste water treatment facility:

<www.epa.gov/cookbook/page90.html>

Local Economy

Community and Localization

Relocalize Now! Getting Ready for Climate Change and the End of Cheap Oil (a Post Carbon Guide), by Julian Darley, Celine Rich and David Room (New Society, 2005).

Going Local: Creating Self-Reliant Communities in a Global Age, by Michael H. Shuman (Routledge, 2000).

"Eight Ways to Beat Wal-Mart," by Albert Norman, *The Nation,* 28 March 1994, at <www.Norfolk-county.com/users/claytons/walmart.html>

Information on local currencies and how to start one:

<http://dmoz.org/Society/Organizations/Local_Currency_Systems/>

Understanding and Creating Alternatives to Legal Tender, by Thomas Greco, Jr. (Chelsea Green, 2002). ☞

Public Power

American Public Power Associates: < www.appanet.org/>

Links to information on energy co-ops:

<www.energy-co-op.net/links.html > and

Community Design

Eco-City Dimensions: Healthy Communities, Healthy Planet, by Mark Roseland (New Society, 2002).

Our Ecological Footprint: Reducing Human Impact on the Earth, by Mathis Wackernagel and William Rees (New Society, 1996).

Information and links on ecocities:

<www.preservenet.com/politics/Ecocities.html> and

Local Governance

Toward Sustainable Communities: Resources for Citizens and Their Governments, by Mark Roseland (New Society, 2002).

The Progressive Launchpad, political and activist links:

<www.geocities.com/Athens/Acropolis/2300/LaunchPad-Politics-sites.html>

Intentional Communities

Ecovillages, by Jan Martin Bang (New Society, 2005).

Creating a Life Together: Practical Tools to Grow Ecovillages and Intentional Communities, by Diana Leafe Christian (New Society, 2003).

The Cohousing Handbook: Building a Place for Community, by Chris ScottHanson and Kelly ScottHanson (New Society, 2004).

Communities magazine: Fellowship for Intentional Community, 138 Twin Oaks Road, Louisa, VA 23093, USA: <http://fic.ic.org/cmag/>

Communities Directory: A Guide to Intentional Communities and Cooperative Living Fellowship for Intentional Community, 2000: <http://directory.ic.org/>

Global Ecovillage Network: < www.gaia.org/>

Dancing Rabbit Ecovillage: < www.dancingrabbit.org/> ∎

intentional communities in existence today, and others in the formative stages, many of which are pioneering a post-industrial lifestyle.

During the energy decline, life in an intentional community could offer many advantages. Association with like-minded people in a context of mutual aid could help overcome many of the challenges that will arise as the larger society undergoes turmoil and reorganization. Moreover, new cooperative, low-energy ways of living can be implemented now, without having to wait for a majority of people in the larger society to awaken to the necessity for change.

However, these advantages do not come without a price: to live successfully in an intentional community requires work and commitment. Many communities fail and many members drop out, for a variety of reasons — often centering on the individuals' projection of unrealistic expectations onto the group. Nevertheless, some communities have managed to survive for decades, and few members of successful communities would willingly trade their way of life for that of the alienated urbanite.

Ecovillages are urban or rural communities that strive to integrate a support-ive social environment with a low-impact way of life. They typically experiment with ecological design, permaculture, natural building materials, consensus decision-making, and alternative energy production. Ecovillages currently scat-tered around the world include the Findhorn Foundation in Scotland; EcoVillage at Ithaca, in Ithaca, New York; the Farm in Summertown, Tennessee; Earthaven in North Carolina; and Mitraniketan in Kerala, India. One that I have visited, the Dancing Rabbit Ecovillage in northeastern Missouri, was established in the early 1990s by a group of young West Coast recent college graduates. Today the community consists of about fifteen adults and children. The buildings were constructed by the residents from natural materials (straw bales and earth), their vehicles all run on vegetable oil, and the community has its own currency and grows most of its food on-site.

The Nation

It is easiest to exert influence on the political process at the local community level. State and provincial politics are almost invariably subject to more compe-tition for access and power, and thus to more corruption. At the national level, the degrees of competition and corruption are truly daunting.

However, many of the legal and economic structures that prevent industrial societies from more quickly and more easily adapting to the energy decline can only be altered or replaced nationally. Even small, incremental changes at this level of government can have important effects. Thus it is essential for citizens who are aware of energy-resource issues to direct at least some of their efforts toward encouraging change at the highest levels of political organization.

Alternative energies and conservation. As nonrenewable energy sources become depleted, it is crucial that renewable substitutes be developed and implemented to replace them — to the degree that replacement is possible. Individuals and local communities can help this happen, but a systematic national policy is badly needed. Policymakers cannot simply wait for the price of nonrenewable sources to rise and that of renewables to fall so that the market itself automatically effects the transition. It will take decades to rebuild the national energy infrastructure, and price signals from the dwindling of nonrenewables will appear far too late to be of any help. In fact, it is already too late to make the transition painlessly. An easy transition might have been possible if the nation had begun the project in the 1970s and continued it consistently and vigorously through to the present. Still, even at this late date, a truly heroic national effort toward developing renewables could succeed in substantially reducing social chaos and human suffering in the decades ahead.

Until recently, the US was spending more on renewable-energy technologies than any other nation, on both a per-capita and an absolute basis. However, Germany, Japan, Spain, Iceland, and the Netherlands are now moving quickly ahead with renewables, while national efforts in the US are stalled.[9]

In view of the absolute dependence of industrial society on energy resources and the imminent decline in fossil fuel availability, one would think that the search for alternatives must be the nation's first priority. Yet the 2004 budget gives its biggest priority instead to the military. The Bush administration's proposed military budget *increase* from 2003 to 2004 is itself larger than the *entire* military budget of any other country in the world except Russia.[10] One cannot help but wonder: Which would be more likely to provide security for us and the next generation — yet another expensive weapons system or a reliable, non-depletable source of energy?

The federal government could and should speed the transition to renewable energy sources by providing substantial tax breaks for individuals who invest in wind and solar, and subsidies for utilities that switch to renewables. Carbon taxes should be implemented and gradually raised — not only to discourage the use of nonrenewables, but also to provide funds for rebuilding the energy infrastructure.

Conservation should also be a high priority: the inability of Congress to pass laws mandating higher auto fuel-efficiency standards is an embarrassment and a disgrace. However, stringent efficiency mandates should be passed not only for automobiles but for a range of appliances and industrial processes. The nation should set goals of reducing the total energy usage by two percent per year, and of progressively altering the ratio of nonrenewable to renewable sources. There may be an economic price for such policies, but it will pale in comparison to the eventual costs of the present course of action.

Food systems. We need to redesign our national food system from the ground up. Currently, that system is centralized around giant agribusiness corporations that control seeds, chemicals, processing, and distribution. Most farmers are economically endangered. We need a national food policy that encourages regional self-sufficiency. This will require a 180-degree shift in how farm subsidies are designed and applied.

Resources for National Policy Change

Alternative Energies

Ron Swenson's comprehensive website on renewable energies:
<www.ecotopia.org>
An excellent general site for news on renewable energies:
Department of Energy, Office of Energy Efficiency and Renewable Energies:
The National Renewable Energy Laboratory: <www.nrel.gov>
The Rocky Mountain Institute: < www.rmi.org>

Food Systems

Farming in Nature's Image: An Ecological Approach to Agriculture, by D. D. Soule and Judy D. Soule, with Jon K. Piper (Island Press, 1992).
Agroecology: Ecological Processes in Sustainable Agriculture, by Stephen R. Gliessman (Lewis, 1998).
Bringing the Food Economy Home: Local Alternatives to Global Agribusiness, by Helena Norberg-Hodge, Todd Merrifield, and Steven Gorelick (Zed, 2002).
Beyond Beef: The Rise and Fall of the Cattle Culture, by Jeremy Rifkin (Dutton, 1992).

Current farm subsidies encourage huge agribusiness corporations and energy dependence. As Philip Lee recently argued in an article in the *Ottawa Citizen*, Canada's agricultural subsidies are similarly promoting centralized, fuel-fed agriculture over sustainable, diversified, local food production.[11]

A range of problems surrounding industrial agriculture could be solved simply by ending current farm subsidies — or, better yet, by instituting an entirely different regime of subsidies that would benefit diversity rather than monocropping; small family farms rather than agribusiness cartels; and organic farming rather than biotech- and petrochemical-based farming.[12]

Financial and business systems. The changes needed in the national economic structure go far beyond efforts to improve accounting regulations so as to avoid more corporate bankruptcies on the scale of Enron or WorldCom. The entire system — designed for an environment of perpetual growth — requires a complete overhaul.

Giant corporations are engines of growth and have become primary power wielders in modern industrial societies. One way to rein them in would be to challenge important legal privileges they have acquired through dubious means. The Fourteenth Amendment to the US Constitution was adopted soon after the Civil War to grant freed slaves the rights of persons; but by the last decades of the 19th century, judges and corporate lawyers had twisted the Amendment's interpretation to regard corporations as persons, thus granting them the same rights as flesh-and-blood human beings. Since then, the Fourteenth Amendment has been invoked to protect corporations' rights roughly 100 times more frequently than African Americans' rights.[13]

The legal fiction of corporate personhood gives corporations the right of free speech, under the First Amendment to the US Constitution. In recent years, when communities or states have sought to restrict corporations' campaign donations to politicians, the courts have overruled such restrictions as a violation of corporate free-speech rights as persons. Corporations also are allowed constitutional protection against illegal search and seizure so that decisions made in corporate boardrooms are protected from public scrutiny. However, corporate "persons" do not have the same limitations and liabilities as flesh-and-blood persons. A human person in California who commits three felonies will be jailed for 25 years to life under that state's "three-strikes" law; but a California-chartered corporate "person" that racks up dozens of felony convictions for breaking environmental or other laws receives only a fine, which it can write off as the cost of doing business. Personhood almost always serves the interests of the largest and wealthiest corporations while small, local businesses that also have corporate legal status are systematically disadvantaged.

Americans should unite behind a national movement to rescind corporate personhood; until that goal is achieved, they should petition state attorneys general to review or revoke the charters of corporations that repeatedly harm their communities or investors.

Our current monetary system, which is based on debt and interest and thereby entails endless economic growth and snowballing indebtedness, requires replacement. While some monetary theorists advocate a gold-based currency as a solution, others argue that a well-regulated, non-debt-based paper or computer-credit currency would have greater flexibility. There is at least one precedent in this regard: the Isle of Guernsey, a British protectorate, has had an interest-free paper currency since 1816, has no public debt, no unemployment, and a high standard of living.[14]

Tax reform is also essential. "Geonomic" tax theorists, who trace their lineage to 19th-century American economist Henry George, argue that society should tax land and other basic resources — the birthright of all — instead of income from labor. Geonomic tax reform, say advocates, could decrease wealth disparities while reducing pollution and discouraging land speculation. Similarly, taxing nonrenewable resources and pollution — instead of giving oil companies huge subsidies in the form of "depletion allowances" — would put the brakes on resource extraction while giving society the means with which to fund the development of renewables.

Population and immigration. Overpopulation is currently one of humanity's greatest problems, and it will become a far greater one with the gradual disappearance of fossil energy resources. But of all the conundrums that beset our species, overpopulation is the most difficult to address from a political standpoint. Both the Left and the Right tend to avoid the issue of continued population growth, or treat it as if it were a benefit. As Russell Hopfenberg and David Pimentel have written,

> [s]ome people believe that for humans to limit their numbers would infringe on their freedom to reproduce. This may be true, but a continued increase in human numbers will infringe on our freedoms from malnutrition, hunger, disease, poverty, and pollution, and on our freedom to enjoy nature and a quality environment.[15]

Currently, the regulation of population is probably best dealt with at a national — as opposed to an individual, community, or global — level because only nations have the ability to offer the incentives and impose the restrictions that will be necessary in order to reverse population growth. The first order of business will be for each nation to gain some sense of its human carrying capacity. Quite simply, if in order to maintain itself a nation is drawing down either nonrenewable resources (such as fossil fuels) or renewables at a faster rate than that at which they can be replaced, then that nation is already over-populated. A cursory scanning of population/resource data would suggest that virtually every nation on Earth has overshot its carrying capacity. This being the case, what should be the target size of national populations? The answer obviously varies from country to country. Globally, according to Hopfenberg and Pimentel,

> [i]f all people are to be fed adequately and equitably, we must have a gradual transition to a global population of 2 billion. A population policy ensuring that each couple produces an average of only 1.5 children would be necessary. If this were implemented, more than 100 years would be required to make the adjustment.[16]

However, this global target needs to be translated into national goals and policies. Again, this is no small challenge.

A frequent tactic in this regard is to appeal to "demographic transition" as an ultimate solution to population problems. In the wealthiest countries, population growth has tended to slow. Germany and Italy, for example, currently have birth rates that are slightly lower than their death rates, which means that their populations are beginning to shrink. This suggests a painless solution to pop-ulation problems: simply increase economic growth in other countries so that they undergo a similar demographic transition to zero or negative population growth. However, as should be clear by now, this is not a realistic option. Industrial growth cannot be maintained much longer even in Europe or North America; much less can we envision fully industrializing all of Africa, South America, and Asia. Another approach must be found.

The empowerment of women within societies also seems to result in reduced population growth. It is women, after all, who give birth and who tradi-tionally provide primary care for young children; given the choice, most women would prefer to bear only a few children and see them grow up healthy and well-fed rather than have many children living in deprivation. Experience also shows that the ready availability of birth control methods and devices is, for obvious reasons, a significant factor in reducing population growth. These strategies, if expanded, will certainly help; however, they probably cannot be counted on to produce the reductions of population size that are actually required in order to avoid famine and public health crises in the coming century.

During the past two decades, China engaged in a unique experiment at population reduction, attempting to limit families to having only one child. Describing this experiment, Garret Hardin writes:

> In some of the major cities the program seemed to be carried forward along the following lines. Decision making was decentralized. Almost every able-bodied woman in a Chinese city was a member of a "production group," which was charged with making its own decisions. Each group was told by the central government what their allotment of rice would be for the year. This allotment would not be readjusted in accordance with the Marxist ideal of "to each according to their need." Rather, it was a flat allotment that made no allowance for increased fertility. It was up to the members of a production group to decide among themselves which women would be allowed to become pregnant during the coming year.[17]

If a member of a production group became pregnant without having obtained permission to do so, she was told to have an abortion.

The results of the Chinese experiment remain unclear since reports reaching the West have been vague and incomplete. There was no doubt a great deal of cheating involved, and farmers and many tribal groups were systematically exempted from the program.

A secondary effect of the Chinese effort occurred in the US, where reports of compulsory abortions in China incited rightist politicians to deny aid to Planned Parenthood and other organizations working to reduce population growth. Many international population programs were consequently seriously undermined by the withdrawal of US participation.

Clearly, the Chinese model — even if it can be said to have been successful in China itself, which is doubtful — will not work everywhere. What other methods are possible?

Ecologist Raymond B. Cowles once suggested using economic motivations to reduce fertility. He proposed simply paying young women *not* to have babies. The expense of such a system would be offset by the savings to society from costs not incurred for education and health care for the children who would otherwise have been born.[18] Economist Kenneth Boulding proposed a somewhat similar *laissez-faire* solution: Instead of money, women should be granted, at birth, a certain number of "baby rights," which could be sold or traded. Lovers of children could buy such rights whereas lovers of money would be encouraged to devote their efforts to activities other than parenthood.[19] Neither Cowles's nor Boulding's idea has garnered much support so far, but both illustrate the kind of creative thinking that must occur if we are to tackle the problem of overpopulation.

Immigration is an issue closely related to that of overpopulation, and it is likewise politically prickly. In the US, roughly 90 percent of the projected population growth for the next 50 years will come from immigration, with the national population projected to double during that time. Such population growth threatens to dramatically increase resource depletion and pollution. From an ecological perspective, immigration is almost never a good idea. Mass immigration simply globalizes the problem of overpopulation. Moreover, it is typically only when people have become indigenous to a particular place after many generations that they develop an appreciation of resource limits.[20]

Opposition to uncontrolled immigration is often confused with anti-immigrant xenophobia. Also, some leftists cogently argue that to cut off immigration to the US from Mexico and other Latin American countries would be unfair: immigrants are only following their resources and wealth northward to the imperial hub that is systematically extracting them. Thus key elements in immigration reduction must be a halt to the US practice of draining wealth and resources from nations to the south, as well as democratization and land reform in the less-consuming countries.

In the decades ahead, all nations must find practical, humane solutions to the problem of population growth and immigration — solutions that will necessarily include legal caps on yearly immigration quotas and some means for reducing both disparities of wealth between nations and the exploitation of one nation by another so that immigration becomes a less attractive option.

As Virginia Abernethy of the Carrying Capacity Network has put it,

> [o]ften, allowing ourselves to be ruled by good-hearted but wrong-headed humanitarian impulses, we encourage ecologically disastrous responses among ourselves and our less fortunate neighbors. Impulses, which seem in the short run to do good, but which lead ultimately

to worldwide disaster — and most quickly to disaster in the countries we wish to help — are not in fact humanitarian.[21]

US Foreign policy. America's military and espionage budgets represent a gargantuan investment in an eventual Armageddon. The US portrays itself as the global cop keeping order in an otherwise chaotic and dangerous world, but in reality America uses its military might primarily to maintain dominance over the world's resources.

This policy is unjust, futile, and dangerous. It is unjust because people in many nations are denied the benefits of their own natural assets. It is futile because the resources in question are limited in extent and their exploitation cannot continue indefinitely and because, by becoming ever more dependent on them, Americans are ensuring their own eventual economic demise. And it is dangerous because it sets an example of violent competition for diminishing resources — an example that other nations are likely to follow, thus leading the whole world into a maelstrom of escalating violence as populations grow and resources become more scarce.

The US policy of maintaining resource dominance is not new. Shortly after World War II, a brutally frank State Department Policy Planning Study authored by George F. Kennan, the American Ambassador to Moscow, noted:

> We have 50 percent of the world's wealth, but only 6.3 percent of its population. In this situation, our real job in the coming period is to devise a pattern of relationships which permit us to maintain this position of disparity. To do so, we have to dispense with all sentimentality ... we should cease thinking about human rights, the raising of living standards and democratisation.[22]

The history of the past five decades would suggest that Kennan's advice was heeded. Today the average US citizen uses five times as much energy as the world average. Even citizens of nations that export oil — such as Venezuela and Iran — use only a small fraction of the energy US citizens use per capita.

The Carter Doctrine, declared in 1980, made it plain that US military might would be applied to the project of dominating the world's oil wealth: henceforth, any hostile effort to impede the flow of Persian Gulf oil would be regarded as an "assault on the vital interests of the United States" and would be "repelled by any means necessary, including military force."[23]

In the past 60 years, the US military and intelligence services have grown to become bureaucracies of unrivaled scope, power, and durability. While the US has not declared war on any nation since 1945, it has nevertheless bombed or invaded a total of 19 countries and stationed troops, or engaged in direct or indirect military action, in dozens of others.[24] During the Cold War, the US military apparatus grew exponentially, ostensibly in response to the threat posed by an archrival: the Soviet Union. But after the end of the Cold War the American military and intelligence establishments did not shrink in scale to any appreciable degree. Rather, their implicit agenda — the protection of global resource interests — emerged as the semi-explicit justification for their continued existence.

With resource hegemony came challenges from nations or sub-national groups opposing that hegemony. But the immensity of US military might ensured that such challenges would be overwhelmingly asymmetrical. US strategists labeled such challenges "terrorism" — a term with a definition malleable enough to be applicable to any threat from any potential enemy, foreign or domestic, while never referring to any violent action on the part of the US, its agents, or its allies.

This policy puts the US on a collision course with the rest of the world. If all-out competition is pursued with the available weapons of awesome power, the result could be the destruction not just of industrial civilization, but of humanity and most of the biosphere.

The alternative is to foster some means of international resource cooperation, but this would require a fundamental change of course for US foreign policy. Daunting though the task may be, it is time to recast US foreign policy from the inside out and from the ground up so that it is based not on resource dominance but on global security through fair and democratic governance structures.

Such a policy shift would necessarily imply both a voluntary relinquishment of US claims on resources and a dramatic scaling back of the US military apparatus. The latter could be accomplished through a fairly swift process of budget cuts, whereby funds formerly devoted to the military would be earmarked instead partly for the dismantling of weapons systems and partly for the redesign of the national energy and transportation infrastructure.

This would have domestic repercussions. Lacking a basis in militarily enforced resource dominance, the US economy would shrink. But this must be seen in perspective: it is an inevitable outcome in any case. The US, as the center of the global industrial empire, does not have the choice of *whether* to decline; it can, however, choose *how* to decline — whether gracefully and peacefully, setting a helpful example for the rest of the world, or petulantly and violently, drawing other nations with it into an accelerating whirlwind of destruction.

Such a unilateral US relinquishment of global dominance would, it could be argued, open the way for another nation — perhaps China — to take center stage. Might Americans wake up one day to find themselves subjects of some alien empire? It may help to remember that the inexorable physics of the energy transition preclude such an occurrence. In the decades ahead, *no* nation will be able to afford to subdue and rule a large, geographically isolated country like the US. Only small, weak, resource-rich nations will be likely targets for conquest.

Transportation. Because of their extreme dependence on car and truck transportation, the US and Canada are, relative to many other industrial societies, at a disadvantage. In the US, the Interstate Highway system represents a vast subsidy to the automobile and trucking industries. Since that system's inception in the 1950s, train transport has languished, with Congress continually reducing the already-small subsidies available for rail transport. In the half-century from 1921 to 1971 (the year of Amtrak's creation), Federal subsidies for highways totaled $71 *billion*; for railroads, $65 *million*. Rail transport has received a total of $30 billion over the past 30 years, whereas federal subsidies for highways in 2002 alone amounted to $32 billion, and for aviation and airports, $14 billion. In the same year, a mere $521 million were set aside for Amtrak. According to a study by the *International Railway Journal*, at $1.64, the US ranks between Bolivia and Turkey in mainline railroad spending per capita. Switzerland spends the most ($228.29) and the Philippines the least ($.29). Urban light-rail systems in the US have fared little better.[25]

America's decades-long shift from rails to highways has been justified by the argument that railroads work better in areas of high population density while highways are more practical where cities and towns are far-flung. Most of the US has a much lower population density than Western Europe and Japan, where rail services move people cheaply and efficiently. In the US, where the distances traveled are typically greater, airlines are more attractive for interurban travel. This argument makes some sense — but only as long as fuel is cheap.

Even with the development of higher-efficiency and alternative-fuel cars and trucks, the energy transition will not permit the continued operation of a national auto/truck fleet of the current size. Moreover, commercial air travel may soon be a thing of the past as jet fuel becomes more scarce and costly. Trains — while still running on fossil fuels — have, when well utilized, lower energy costs per passenger-mile than either cars or planes. Thus, for the US, one sensible course of action would be to immediately cease subsidizing highways and airlines and to begin investing in rails.

At the same time, auto companies would be well advised to put in high gear their research into smaller, lighter, more energy-efficient electric, hybrid, and even human-powered vehicles.

Ultimately, for people in industrial societies, the future holds less travel in store, regardless of the means of transport chosen. Economic survival will thus require reducing the *need* for transportation by moving producers, workers, and consumers closer together.

Activism. In order to feel that the sacrifices they are making during the energy downturn are fair, the people of any nation must be empowered to participate in the process of making decisions about how those sacrifices are allocated. However, the fundamental changes to national economies and infrastructures described above are not likely to be implemented through conventional political means — by citizens voting for candidates — because it is in the interests of most politicians to lie rather than to convey bad news. More people will tend to vote for the candidate who promises the rosier future, even if those promises are patently unrealistic. Therefore the radical shifts needed can probably only happen as a result of the dramatically increased involvement of an informed citizenry at every level of a revitalized political process.

Unfortunately, the citizenry is currently neither informed nor involved, and the system resists fundamental change at all levels. Immense sums are invested annually to distract the public from substantive issues and to turn their attention instead toward consumption and complacency.

The small minority who are aware of the difficult choices facing society need to take heart and redouble their efforts to educate others, including government officials.

Activists could play a crucial role in the upcoming energy transition, as they have played in most of the important social advances of the past few decades. Activist-led social movements have helped end colonialism and the worst manifestations of racism, gained rights for women, and helped protect numerous species and sites of biodiversity. Today many activists are advocating a rapid transition to renewable energy sources, conservation, and the equitable distribution of resources. Moreover, they are leading the way in modeling nonviolent social change.

An example of the latter is Marshall Rosenberg, whose Center for Nonviolent Communication works internationally with such groups as educators, managers, military officers, prisoners, police and prison officials, clergy, government officials, and individual families. Nonviolent Communication trainings evolved from Rosenberg's quest, during the civil rights movement in the 1960s, to find a way of rapidly disseminating much-needed peacemaking skills. Today he is active in war-torn areas (Israel, Palestine, Bosnia, Columbia, Rwanda, Burundi, Nigeria, Serbia, and Croatia), promoting reconciliation and a peaceful resolution of differences. As social and economic pressures from the energy transition mount, such mediation efforts could — both globally and locally — mean the difference between peaceful cooperation and savage competition.

Social activists tend to be the leading-edge thinkers and change agents for society as a whole. We need more of them.

The World

Many people are wary of world government, believing that it would lead inevitably to global tyranny. This fear is both founded and unfounded. It is well-founded in the sense that people's individual ability to contribute to the decisions that affect their lives varies inversely with the scale of social organization: it is easier to make one's voice heard in a town meeting than in a national election. This being the case, it seems highly likely that a world government, were one ever to be established, would tend to be remote and unresponsive to the needs of individuals and local communities. But the fear is unfounded in the sense that, without fossil fuels, it is doubtful that a sufficient energy basis could ever be assembled to build and maintain a government with a global scale of organization, communication, and enforcement.

Hence the reasonableness of a principle succinctly stated by ecologist Garrett Hardin: *Never globalize a problem if it can possibly be dealt with locally.*[26]

Are there any problems that *must* be dealt with globally? In ordinary times, there probably are not. However, during the extraordinary period of the peaking and decline of fuel-based industrialism in which we are now living, there are three kinds of problems that do indeed demand some kind of global regulatory mechanism: resource conservation, large-scale pollution control, and the resolution of conflicts between nations. All three must be *administered* more or less locally: Resources exist in geographically circumscribed areas that are ultimately the responsibility of regional decision-making bodies; pollution often issues from point sources that are best monitored by local agencies; and conflicts must ultimately be resolved by the parties involved. But the depletion of internationally traded essential resources, industrial production processes, and industrialized warfare are capable of having overwhelming global effects. Catastrophic global warming and nuclear war provide compelling examples: either would result from decisions made, and actions taken, by specific people in particular places; but the consequences of those decisions and actions would profoundly impact people and other organisms everywhere. The consequences are so far out of proportion to the decisions and actions taken locally that some form of global control mechanism seems called for, consisting of enforceable minimum conservation standards and enforceable means of containing or resolving international conflicts.

Some agencies already exist for addressing global problems. They are generally of three kinds: first, corporations, trade bodies, and lending institutions; second, the quasi-governmental apparatus of the UN, with its related aid agencies; and third, the small but vocal cadre of transnational human rights and environmental NGOs.

The corporations, international banks, and trade bodies together constitute a force for globalization-from-above. They are doing almost nothing to help, and much to hinder, an orderly global energy transition. This should be no surprise: they are part and parcel of the growth economy that flows from the fossil-fuel pipeline.

The forces of globalization-from-below (the NGOs) do not have a full picture of the degree to which world events revolve around energy resources and their depletion; nor do they have an adequate strategy for dealing with the issues they are confronting. But their push toward decentralization, democratization, and cooperation is nevertheless generally the right way to help humanity wean itself as painlessly as possible from fuel-fed industrialism. Thus what is needed globally is a weakening of the forces of globalization-from-above and a strengthening of those of globalization-from-below.

The UN — which is caught somewhere between those two sets of forces — is one of the few institutions that is in any position to provide enforceable minimum global environmental standards and to serve as an arena for conflict resolution.

If all parties concerned understood the severity of the crisis facing them, there is much they could do. They could negotiate more global agreements modeled on the Kyoto accords, ensuring international efforts to reduce greenhouse gas emissions and subsidize renewables. The International Energy Agency could be expanded and empowered to survey, conserve, and allocate energy resources in such a way that all nations would have assured (though diminishing) access to them, and that profits from resource exploitation would go toward helping societies with the transition, rather than merely further enriching corporate executives. Meanwhile, UN-based conflict-resolution and weapons-destruction programs could substantially reduce the likelihood of violent conflicts erupting over resource disputes. Rich industrialized nations could wean themselves as quickly as possible from fossil fuels while less-industrialized nations, abandoning the futile effort to industrialize, could embark on the path of truly sustainable development. Industrialized nations could assist the latter in doing so by ceasing the practice of siphoning off less-consuming nations' resources.

What is especially needed is a new global protocol by which oil-importing nations would agree to diminish their imports at the rate of world depletion — approximately two percent per year. That way, price swings would be moderated as the peak of global oil production passes, enabling poorer nations to be able to continue importing the bare minimum of resources needed to maintain their economies. A Model for such an agreement has been proposed by the Association for the Study of Peak Oil (ASPO)

The majority of the world's nations and peoples would probably be willing to participate in all of these difficult and even painful efforts if they were informed clearly of the alternatives. The greatest impediment would likely be the non-participation of a few "rogue states" that tend to disregard international laws and treaties at will. The foremost of these are the US and, to a lesser extent, China.

With only five percent of the world's population, the US has the lion's share of the world's weaponry and exercises direct or indirect control over a steeply disproportionate share of global resources. The US cleans up some pollution at home while undermining international environmental agreements. It refuses international inspection of its weapons of mass destruction and attacks other nations virtually at will. It also undermines efforts to stabilize or reduce the global population at every turn. Just within the past four years, the US has abrogated the anti-ballistic missile treaty and undermined the small arms treaty, the UN convention against torture, the international criminal court, and the biological weapons convention.

Will the US join the international community, or insist on maintaining its privileged status even as its empire crumbles? This is the first great geopolitical question we face as the industrial interval wanes.

The second one concerns China. Will China continue to seek to industrialize? Because of its huge population, efforts in that direction will put great stress on any global efforts at conservation and pollution abatement.

The world does not revolve around these two countries. But if they could be persuaded — by either their own citizenry or the international community — to exercise constructive leadership, the global energy transition could occur far more smoothly than would otherwise be the case.

Taken together, these recommendations imply a nearly complete redesign of the human project. They describe a fundamental change of direction — from the larger, faster, and more centralized to the smaller, slower, and more locally based; from competition to cooperation; and from boundless growth to self-limitation.

If such recommendations were taken seriously, they could lead to a world a century from now with fewer people using less energy per capita, all of it from renewable sources, while enjoying a quality of life that the typical industrial urbanite of today would perhaps envy. Human inventiveness could be put to the task of discovering ways not to use more resources, but to expand artistic satisfaction, find just and convivial social arrangements, and deepen the spiritual experience of being human. Living in smaller communities, people would enjoy having more control over their lives. Traveling less, they would have more of a sense of place and of rootedness, and more of a feeling of being at home in the natural world. Renewable energy sources would provide some conveniences, but not nearly on the scale of fossil-fueled industrialism.

This will not, however, be an automatic outcome of the energy decline. Such a happy result can only come about through considerable conscious effort. It is easy to imagine less desirable scenarios.

There are, at the local level, many hopeful signs that a shift toward sustainability is beginning. But there are also many discouraging signs that large political and economic institutions will resist change in that direction. Seeing the latter signs and the immensity of the challenge before us, we can easily drift into discouragement and inaction. Is it too late? Are recommendations for a peaceful energy transition hopelessly unrealistic?

In some respects, it *is* too late. As noted above, the transition could have been made much more easily if we had started 30 years ago, and with a World War II-level of effort. Every few years since the oil crisis of 1973, another book has been published that says, in essence, "We have little time left; we must start now to change direction before it is too late." At the Earth Summit in Rio de Janeiro in 1992, several eminent speakers agreed that the global community had the decade of the '90s in which to turn from growth and consumption toward sustainability. The turn was not made in that decade. Indeed, the tread-mill of consumption only accelerated. At what point does the clock run out? Is there a time when we'll have to say, "We had our chance and blew it"?

If by "Is it too late?" we mean "Is it too late to make the transition painlessly?", then the answer may well be *yes*. By now, we almost certainly face a "disconti-nuity," as renewable-energy expert Ron Swenson euphemistically put it in a recent phone conversation with me.

Am I being fatalistic? Or simply realistic? Our cultural obsession with good news, promises, and hope is humanly understandable, but there comes a time when the best thing to do is to accept that a bad situation has developed and to find intelligent ways to manage it.

If by "Is it too late?" we mean "Can we do anything now to make the future better than it would otherwise be?", then the answer, of course, is that it is never too late. There is much to be done, and hard work now may yield great benefits for the generations to follow. We may have up to 10 years before the gross level of energy available from all hydrocarbon resources falls signifi-cantly below current figures (though the net level of available energy may decline much sooner). Much can be done during that time. However, we need to acknowledge that waiting has consequences. The more we do, and the sooner we start, the better off we will be.

Are these recommendations for national and global change unrealistic? Past experience would suggest that national leaders will be unlikely to act on the basis of warnings like those contained in this book. I have already explored to some extent the reasons for their reluctance: a political system based on mon-eyed influence and a monetary system based on debt and interest, and hence endless growth. Thus, in order for many of the above recommendations to be implemented, much more may be involved than the technical problem of replacing the energy infrastructure of industrial societies (which is, of course, no small feat in itself). In order for that latter task to begin in earnest, we will also probably have to make fundamental changes to both our political systems and our economic systems. A successful transformation of even one of these three aspects of any single industrial society — its energy infrastructure, its political system, or its economic system — represents a daunting task probably requiring decades of work by many thousands of people. The likelihood of achieving fundamental change in all three arenas, and in most industrial nations, before the repercussions of the energy decline are felt is surely remote.

Nevertheless, the proposals need to be on the table. The public needs to know that there are alternatives to continued growth, resource competition, and chaotic collapse. The mere fact that nearly every one of the above proposals is being put forward by many individuals and groups in many places suggests that there is already a growing awareness that we cannot keep going much further in the direction in which we are now headed. If the leaders cannot lead, they must get out of the way and let the people make the needed changes.

Moreover, while proposals for basic infrastructural, economic, and political change may seem hopelessly unrealistic within the current context, we must remember that the context is shifting. Times of crisis offer both danger and opportunity, and we are approaching a time of cascading crises — and hence, perhaps, large and unexpected opportunities.

A Final Word

I would like to close with some personal observations. My experience of writing this book has been somewhat distressing at times, for reasons that should be fairly obvious. The subject I have chosen is not particularly cheery, a fact I have underscored in the book's title and subtitle. Surely the reader's engagement with this material has also brought an occasional moment of mental unease.

However, writing this book has also been rewarding in several respects, and so it seems important that I point out these rewards, if only to reassure readers that their task may also have been worthwhile.

First, I believe I have gained from this study *a better understanding of many of the most important problems now facing humanity, and of those likely to do so in the foreseeable future.* We all see daily evidence that the world is an increasingly unsettling, dangerous place: every morning's newspaper is likely to inform us of some new battle, terrorist act, or economic disaster. What are we to make of it all? Is it the work of Satan? Are foreign despots or greedy corporate executives to blame? Are we victims of the wrongheaded schemes of liberals and socialists? Is a vast conspiracy afoot?

An investigation of the history of humanity's evolving relationship with energy resources suggests a prosaic explanation. The growing turmoil we see around us is primarily the inevitable result of a way of life our immediate ancestors adopted for reasons that seemed to them self-evidently compelling, a way of life that we have accepted as inevitable and ordinary. Human beings have always had problems: competition for scarce resources, natural disasters, diseases, accidents, and so on. It is the *scale* of the problems that beset us now that is unique. The steep expansion in scale of the human population size and the consumption of resources that has characterized modern societies is almost entirely due to industrialism and the use of fossil fuels. And many of the large-scale problems that we are likely to encounter in this century will be due to the depletion of those fuels.

When we operate on the basis of false explanations, we live in a state of confusion, and our attempts to solve problems are unlikely to be effective. With a more accurate understanding of problems, we have a much better chance of addressing them successfully. We also are more likely to place blame where it belongs, if indeed blame is called for.

Explanations for social problems usually carry moral implications, and the explanation offered in this book is no exception. We like to think that our human intelligence and our moral codes set us apart from other organisms. When other creatures gain an energy subsidy, they instinctively react by proliferating: their population goes through the well-studied stages of bloom, overshoot, and die-off. If we humans are more than mere animals, we should be expected to behave differently. Yet so far we have reacted to the energy subsidy of fossil fuels exactly the way rats, fruit flies, or bacteria respond to an abundant new food source. A hard look at the evidence tends to make one skeptical of human claims of specialness, causing one, almost inevitably, to view more sympathetically the choice our species has made to become socially dependent on nonrenewable fuels.

Of course, throughout the period of industrialization, matters could have been handled better at every stage. This is putting it charitably. The decades of the Industrial Revolution were replete with outrages against humanity and nature (many of which have been documented by Derek Jensen in his brilliant and harrowing book *The Culture of Make Believe* [27]). Unquestionably, a relatively few people and institutions have been responsible for immense suffering. While this was also the case throughout the millennia before the Industrial Revolution, industrialism and fossil energy resources, by vastly expanding the power wielded by human beings, also vastly expanded the human ability to commit atrocities. That being the case, perhaps it makes good moral sense to keep the scale of our societies and their projects small in the future so that the crimes and outrages that human beings will inevitably continue to commit will also occur on a small scale.

In the final analysis, we are left with many of the same moral questions as always, but we see those questions in a new light. The end of industrialism may lead us to be both more critical and at the same time more understanding of human foibles — more critical because we see writ large the results of greed and unrestrained competition; more understanding because it is clear that we humans are, at least to a very large degree, simply animals responding to biological urges and environmental circumstances. Our vaunted moral and intellectual capabilities may enable us to alter our behavior, but perhaps only within rather narrow limits. What those limits are remains to be seen. If ever we have had an opportunity to prove our specialness as a species, our ability to collectively exert moral and intellectual faculties to overcome genetic programming and environmental conditioning through intelligent self-limitation, it is now.

A second reward I have gleaned from writing this book has been *a more realistic sense of the human goals that are achievable under the present circumstances.* Unrealistic goals breed disappointment and disillusionment. If we expect for our children the kind of high-energy society that we ourselves have known, our hopes will be dashed. We thereby set ourselves up for continual disappointment. As I have said, I believe that it is realistic to hope for a future world of smaller, more egalitarian communities in which people have more time on their hands and live closer to nature. It is realistic to hope for humankind to move collectively from being a colonizing species to being a cooperative member of climax ecosystems. Other species do this frequently, and various human cultures have made such a transition in the past. With effort, we can achieve this goal with minimal human suffering and environmental destruction in the interim.

However, even after we have downsized our long-term vision for society, we may still be frustrated because we don't see quick progress toward that goal. We humans like quick results. But sometimes people live in times when things aren't getting easier and when their efforts toward building a better world seem to bear little fruit. What should one's attitude be if one is living in such a time? How does one continue to invest effort toward making positive change in the world without succumbing to cynicism and burnout? Surely it is helpful, during such times, to have an overarching historical perspective so that one has the sense of contributing to an eventual desirable outcome *which one may not be around personally to see.*

Third and finally, I believe I have gained *a heightened sense of my generation's responsibility.* Those of us who are older adults (I'm now 54) have lived in the most exciting time in history. Even if we have suffered from the stresses of the fast pace, the pollution, and the economic competition of modern life, we have benefited from the enormous energies at our disposal. Most of us — most, at least, who have grown up in industrialized countries — have lived free from hunger, with hot and cold running water, with machines at our fingertips to transport us quickly and almost effortlessly from place to place, and with still other machines to clean our clothes, to entertain and inform us, and on and on.

It has been a fabulous party. But from those to whom much has been given, much should be expected. Once we are aware of the choice, it is up to us to decide: Shall we vainly continue reveling until the bitter end, and take most of the rest of the world down with us? Or shall we acknowledge that the party is over, clean up after ourselves, and make way for those who will come after us?

Afterword to the Revised Edition

I n the two years since the publication of the original edition of *The Party's Over*, the discussion of the phenomenon of peak oil and the economic and geopolitical turmoil likely to arise from it has moved from the fringe to the mainstream. Over half a dozen other books on the subject of the limits to the production of fossil fuels have appeared — including Julian Darley's *High Noon for Natural Gas*, Paul Roberts' *The End of Oil*, David Goodstein's *Out of Gas*, Sonia Shah's *Crude*, and Dale Allen Pfeiffer's *The End of the Age of Oil*. At least three organizations have been formed to research the problem of oil depletion and possible responses, including the Association for the Study of Peak Oil (ASPO); the Oil Depletion Analysis Centre (ODAC); and the Post-Carbon Institute (PCI). Additionally, a documentary film, "The End of Suburbia," (<www.endofsuburbia.com>), centering on the potential impacts of peak oil on the American way of life, and has created a minor underground sensation.

Numerous relevant websites have also sprung up, including <lifeaftertheoilcrash.net>, <energybulletin.net>, <peakoil.org>, and <oilcrisis.com>.

Soaring oil prices during 2004 prompted headlines in the *New York Times* ("The Oil Crunch," by Paul Krugman, and a May 19, 2004 editorial titled "Gasoline Hysteria"), *Le Monde* ("The Petro-Apocalypse," by Yves Cochet), CBS *Marketwatch* ("The Looming Oil Crisis Will Dwarf 1973," by Paul Erdman), and elsewhere. Even *National Geographic*, in its June 2004 cover story, proclaimed "The End of Cheap Oil."

In the short term, high oil prices appeared to be due to increased demand, lack of refining capacity in the US, and instability in the Middle East (Iraq's production just can't seem to get off the ground, due to repeated efforts at sabotage on the part of the indigenous population, and reluctance on the part of the oil companies to invest there, given the unsafe working conditions). "So why wouldn't oil prices rocket?" asked Alan Kohler in the title of his May 19, 2004 essay at <www.smh.com.au/articles/2004/05/18/1084783514440.html>. "The last 'super giant' oilfield (more than 10 billion barrels) was discovered 40 years ago; the last American refinery was built 25 years ago; each successive American 'driving season' guzzles more gas than the last."

Although major daily newspapers talked about the immediate causes of high gas prices, they only occasionally noted that these were riding on a deeper, tectonically shifting terrain.

The Saudi Enigma

Global spare production capacity (the amount that exporting nations *could* produce if called upon, over and above what they are now producing), is now at its lowest point in recent decades — reportedly a mere 1 to 2 million barrels per day out of a total global output of about 83 million barrels per day. And most of the spare capacity exists in one nation — Saudi Arabia. But even this assessment, worrisome as it may be, rests on the assumption that official Saudi reserve estimates are correct.

As mentioned in Chapter 3, for the past three years oil investment banker Matthew Simmons has been publicly questioning whether Saudi oil wells really contain all of the oil that Saudi officials claim is there. In articles in the *New York Times* and is his new book, *Twilight in the Desert: The Coming Saudi Oil Shock and the World Economy*, Simmons has been quoted as saying that his extensive review of 200 technical papers by scientists working in the Saudi fields has led him to doubt the published figures. For many years, the country's five major oil fields — including Ghawar, the largest oil field ever discovered — have provided the core of Saudi production, but oil field operators are injecting millions of barrels of sea water each day in order to maintain pressure within the underground systems. This practice maintains extraction levels; however, the aging Saudi fields — all discovered between 1940 and 1965 — are inevitably being depleted. When the inevitable decline in extraction rates begins, seawater injection could actually accelerate the process, resulting in a rapid drop-off in oil available for the export market.

Simmons's statements were evidently so worrisome to Saudi officials that the latter arranged a high-profile symposium at the Center for Strategic and International Studies in Washington, DC in late April 2004. Their own representatives, together with prominent US government officials, assured the world that Saudi Arabia's oil fields are robust and able to supply increasing global petroleum demands for decades to come. Saudi officials even took the extraordinary step of announcing that official reserve estimates of 261 billion barrels of recoverable oil are far too *low*. For this claim to be credible, however, independent analysts will have to see credible evidence of spectacular new discoveries — of which no word has yet leaked out. Unless such evidence emerges, it would probably be safe to characterize the Saudi statements as an act of desperation intended to shore up US support for the increasingly embattled monarchy.

In October 2004, Channel 4 News in Britain conducted an interview with Sadad Al Husseini, the recently retired vice-president for exploration of the Saudi oil company Aramco. In the interview, Husseini noted that official US forecasts for future oil supplies (which assume that Saudi Arabia can expand its oil production by over 100 percent over the next two decades), are a "dangerous over-estimate." Asked if people should be worried by the actual state of affairs, he replied in the affirmative.

Given the context of recent events, Mr. Husseini's comments carry considerable significance. They represent a radical break from previous Saudi official statements and signal that the nation with the world's largest stated petroleum reserves cannot, in fact, continue to open the oil spigot arbitrarily in order to keep prices low.[1]

Shell Game

Meanwhile, in the spring of 2004, Royal Dutch/Shell created shock waves by reducing its reported reserves on three separate occasions within a nine-week period. This 20 percent total reserve reduction was startling enough, but an examination of the reasons for the embarrassing corporate admission (which resulted in the firing or resignation of several high-level executives and the hammering of Shell stock prices), leads to even deeper questions about standard industry reporting practices, and about technologies that are being relied upon to extend current oil production levels in many countries.

Many of Shell's difficulties issued from the oil-exporting nation of Oman, where production levels have been declining for the past four years. Shell executives in that country apparently expected that horizontal drilling techniques would be able to maintain and even increase extraction rates. These expectations led them to overestimate their company's reserves within that nation by as much as 40 percent. A similar situation in Nigeria also led to downward reserve revisions.

This was bad enough for Shell, but the really grim news is what is implied for the rest of the industry. Other companies active in Nigeria — including

Italy's ENI, France's Total, and US-based ChevronTexaco and ExxonMobil — appear to have followed Shell's practice of exaggerating reserves. While new technologies — which many oil optimists are relying on to fulfill rosy projections for increased global production — appear to be effective at extracting oil from known reserves more quickly and efficiently, the overall result seems to be simply the quicker exhaustion of those reserves.

Oil's Depressing Outlook

Even as questions are being raised about global oil supply, demand is inexorably growing. China is currently increasing its oil imports by 30 percent per year, and in 2003 that nation surpassed Japan to become the world's second foremost petroleum importer. In the same year, Shanghai banned bicycles from most of its main streets in favor of automobiles.

As Chris Skrebowski of *Petroleum Review* notes in his November 2004 report "Oil Field Megaprojects," several substantial deepwater oil fields are scheduled to come on-stream in 2006, so there is some possibility of a stabilization of prices. Moreover, if current high prices lead to a renewed global recession, this could result in a drop in demand, which could in turn lead to lower fuel prices. But that effect would only be temporary. From the long-term perspective, burgeoning demand is on a collision course with emerging supply constraints, and $60, $80, and even $100 per barrel oil is possible in the near term.

When will the actual global peak of oil production occur? In the original edition of *The Party's Over*, I surveyed several authoritative forecasts and, on that basis, cited a decade-long window of 2006 to 2016 as the most likely period during which the global all-time peak in oil production will take place. The latest data — from *Petroleum Review* and Matthew Simmons, among other sources — suggest that the peak may more likely occur during the earlier years of that window. Between now and then, we will continue to experience a bumpy ride as we leave the "petroleum plateau" that we have been on for the past 30 years. Once we are off the plateau and on the downward skid, times may get very interesting indeed.

Significant New Reports

During the first months of 2005, several reports relevant to the issue of peak oil were issued; each had important implications that can only be summarized briefly here.

The Hirsch Report. Commissioned by the US Department of Energy from Science Applications International Corporation (SAIC) and released in February, the study titled "Peaking of World Oil Production: Impacts, Mitigation and Risk Management," led by Robert L. Hirsch, examines the likely consequences of the impending global peak. The Executive Summary begins with the following paragraph:

> The peaking of world oil production presents the U.S. and the world with an unprecedented risk management problem. As peaking is approached, liquid fuel prices and price volatility will increase dramatically, and, without timely mitigation, the economic, social, and political costs will be unprecedented. Viable mitigation options exist on both the supply and demand sides, but to have substantial impact, they must be initiated more than a decade in advance of peaking.

The report offers three scenarios: one in which mitigation efforts are not undertaken until global oil production peaks; a second in which efforts commence only ten years in advance of peak; and a third in which efforts begin twenty years prior to the peak. Each scenario assumes a "crash program rate of implementation." In the first case, the study suggests that peak will leave the world with a "significant liquid fuels deficit for more than two decades" that "will almost certainly cause major economic upheaval"; even with a ten-year lead time for mitigation efforts government intervention will be required and the world will experience a ten-year fuel shortfall. A crash program initiated twenty years ahead of the event will off "the possibility" of avoiding a fuel shortfall. The report emphasizes repeatedly that both supply- and demand-side mitigation options will take many years to implement; it also notes that "The world has never faced a problem like this."

The IEA Report: "Saving Oil in a Hurry: Measures for Rapid Demand Restraint in Transport." The International Energy Agency has released, in draft form, a small book advising countries to prepare contingency plans to be implemented in the case of petroleum supply shortfalls. While not specifically predicting such shortfalls, the book analyzes the supply disruptions of the 1970s to see which demand-restriction measures were most helpful. The report advises developing policies such as:

- Driving bans on alternate days (if your license plate ends with an odd number, you would be allowed to drive on Mondays; Wednesdays, and Fridays; if it ends with an even number, you could drive on Tuesdays, Thursdays, and Saturdays).

- Reduced speed limits
- Encouragement of telecommuting
- A 50% reduction in public transport fares
- Building more carpool lanes, and making existing ones active on a 24-hour basis

The Bank of Montreal Report: "Big Footprints on the Sands of Time, and Little Footprints of Fear." In the course of this report, released March 30, 2005 by Harris Investment Management, Inc. (a member of the Bank of Montreal Investment Group), author Donald G. M. Coxe notes that even newly developed oil fields in Saudi Arabia are being pressurized with desalinated water from the Arabian Gulf. "Isn't waterflooding petroleum Viagra for aging wells?," asks Coxe. He goes on to speculate that

> the combination of the news that there's no new Saudi Light coming on stream for the next seven years plus the 27% projected decline from existing fields means Hubbert's Peak has arrived in Saudi Arabia. The Kingdom's decline rate will be among the world's fastest as this decade wanes. Most importantly, Hubbert's Peak must have arrived for Ghawar, the world's biggest oilfield, and Wall Street's most-cited reason for assuring us month after month that oil prices would plunge because there were so many billions of barrels of readily-available crude overhanging the market.

The report goes on to say that news from Mexico's Canterell, the world's second-largest field, and from the North Sea as well, is just as bad, and concludes that "oil shortages are here to stay."

The Goldman Sachs Report. This report, issued March 30, does not discuss Peak Oil per se; instead, it warns of an oil price "super spike" period — "a multi-year trading band of oil prices high enough to meaningfully reduce energy consumption" — resulting from surging demand in China and the US. The report suggests that oil prices could hit $105 per barrel by 2007. It also notes that "our new range [$50 – 105 per barrel] could prove conservative, especially if there is a supply disruption in a major oil exporting country.

The Iraq Quagmire

By far the most discussed development since April 2003 (when *The Party's Over* hit the bookstores), has been the US-British invasion and ongoing occupation of Iraq. As I discussed in my subsequent book *Powerdown* (New Society,

2004), I do not believe that this invasion was undertaken simply to commandeer Iraq's oil supply: the situation is more complex, and hinges on the
Washington neoconservatives' published fantasies of world domination.
However, when the Iraq adventure is seen in light of America's long-term foreign policy in the Middle East, it can certainly be regarded as an oil war. The
US would have little interest in that part of the world were it not for the fact
that 60 percent of proven global oil reserves are concentrated there. No doubt
the strategy behind the war included the building of several large and permanent military bases in Iraq for the defense of US access to oil supplies in the
region, especially in neighboring Saudi Arabia.

Accusations that the invasion was motivated by a thirst for oil gained cred-
ibility when American troops, as they entered Baghdad, faithfully guarded the
Iraqi oil ministry but allowed other government buildings — including museums — to be looted. However, despite keen attention on the part of US
civilian contractors, Iraq's oil production has languished, partly due to ongoing sabotage by Iraqi resistance fighters.

By now it is clear that the invasion and subsequent occupation were fraught
with almost unfathomable incompetence and poor planning, all issuing from
arrogant Washington neoconservative ideologues.

Revelations about the torture of Iraqis in American-run prisons have dramatically intensified the widespread perception that the entire exercise was
criminal in nature. Even in the US itself, sentiment is growing that the country
has allowed itself to be taken over by a ring of gangsters who have undermined
the nation's international standing and strategic interests. America now faces a
no-win situation regardless of whether it tries to continue the occupation or
picks up and leaves. In either case it has lost face, made enemies, and squan-
dered opportunities. The entire Middle East has been destabilized, and the
flames of Islamic fundamentalism have been fanned to white heat.

For the world as a whole, the consequences of the Iraq fiasco are likely to
be severe and long lasting. The invasion has created a widening rift between
the US and many other nations. It has also hastened the inevitable energy crisis
(by at least temporarily undermining Iraq's production capacity) and has likely
made that crisis much harder to solve. This is because the destabilization of the
Middle East will lead to greater geopolitical competition for control of
resources. The region cannot simply be left to sort out its problems on its own:
all of the world's oil-importing nations have a survival stake in the contest.
And that contest is likely to become more chaotic in years ahead, as the Saudis

attempt to deal with simmering internal conflicts — an increasing population of younger people, declining per-capita incomes, increasing Islamic fundamentalist sentiment and violence, and ambiguity regarding a successor to the ailing King Fahd.

The old order in the Middle East is nearly finished, and a new one must be negotiated, with the US, Israel, China, Russia, Japan, India, Europe, and the Middle Eastern exporting countries themselves as the primary interested parties. But "negotiated" may be too tidy a term for what lies ahead in the region.

Russia, China, Europe, and Brazil are seeking a "multipolar" world order to replace the American-led regime of corporate globalization that has characterized the period since the end of the Cold War. Meanwhile former US subordinates such as Venezuela, Bolivia, and Argentina have rebelled against the "Washington consensus." The end is in sight for US-led corporate globalization, despite the continuing growth in global trade and the accelerating outsourcing of jobs from the US to India and China.

Largely as a result of the neoconservatives' unbounded hubris, the US economy and geopolitical status are unraveling more quickly than could have been imagined only a few years ago. While in 2004 the US appeared to be in the early stages of an economic recovery, that recovery is being undermined by high oil prices, staggering levels of government debt, and a ballooning trade deficit fed largely by the need for ever greater fuel imports.

The only chance for a peaceful solution to the global energy crisis will be to foster cooperation between nations, the conservation of remaining resources, and the sharing of what oil is left. This is a politically challenging scenario at best, and it has been made far more so by the Bush administration's crimes and blunders.

The Curse of Free Energy

I have received hundreds of messages in response to *The Party's Over*, scores of them suggesting that I have overlooked or underestimated various alternative energy sources. This was certainly the case in at least some instances, and information I have received from readers is reflected in the updated assessments of non-petroleum energy sources contained in Chapter 5. However, the subtext of many of these messages was that alternative energy sources will be capable of sustaining industrial civilization in more or less its present configuration far into the future. With this I disagree.

As I have pointed out in *Powerdown*, it is a mistake to view oil depletion as a technical problem that can be solved by substituting other energy sources for petroleum. This statement may seem counter-intuitive, since to most people it must appear obvious that if we are about to run out of cheap energy, the solution is to find other sources of cheap energy.

The search for supply-side solutions to the problem of resource depletion is time-honored: we humans have become masters at every imaginable strategy for increasing our rates of extraction of important raw materials. The supply-expansion gambit has sometimes succeeded for us spectacularly — as documented in Chapters 1 and 2 of this book. As I also sought to point out there, the effort has not always paid off so well — witness the legacies of civilizations that collapsed because of the depletion of topsoil, forests, grazing lands, or other essential resources. As Joseph Tainter has shown, returns from investments in complexity (which are also, in effect, investments in supply-side strategies) have a tendency to diminish over time.

Nevertheless, the motive for growth is so strong that it leads to a kind of mystique, which takes its ultimate form in what could be termed the cult of the inventor-savior. The cultic myth goes something like this: Once upon a time, the world teetered on the brink of chaos. Society had become mired in inefficient ways of producing or delivering its essential goods. All would have been lost but for the intervention of the Hero — who, through the tireless exercise of his superior intellect, produced an Invention that not only averted calamity but led to the dawn of a new and better era. Thomas Edison and Alexander Graham Bell were among the early inventor-heroes; Nikola Tesla, whose career is discussed in Chapter 2, seems to be the patron saint of the modern "free-energy" branch of the cult.

No one doubts that good ideas are helpful. Better designs and new inventions can indeed, in some instances at least, enable us to do the things we need to do in a more efficient and less wasteful manner. But will technology by itself, or a supply of new resources, or a way of more cleverly extracting or using current resources do anything more than buy us a little time?

Not all cult devotees are so bold as to suggest it, but surely the ultimate dream of those who advocate a technological fix must be some form of free energy. Suppose an inventor-savior were to come up with a simple device that, when operating, actually produced more energy than it consumed. What would be the implications? If the cultic myth is to be believed, it might mean the liberation of humanity from its age-old material burdens; we might therefore experience a collective spiritual awakening. Wars for control of scarce resources might cease. It might mean an end not just to drudgery, but to all forms of poverty and human exploitation — truly, Paradise at last regained!

As enchanting as this mythic vision may be, I contend that it has little to do with reality. In fact, we have had an energy source that was virtually free for the past century. I am speaking not of an exotic perpetual-motion machine based on the ingenious arrangement of permanent magnets, but of ordinary old petroleum. The energy in a single gallon of gasoline is roughly equivalent to the energy expended by a human being working hard (producing a quarter of a horsepower) for a month, and an American working at a minimum-wage job can purchase a gallon of gasoline for about 20 minutes of labor. This is a ratio of 600 to 1. The only monetary investment that I can think of that has a similar rate of return is a winning lottery ticket. Thus, even for a low-wage employee, energy has been and is still so extraordinarily cheap as to be virtually free. Hence our ability to run a society in which the average person has hundreds of "energy slaves." This is probably about as close to truly free energy as human beings will ever get.

And what have we done with this effortless and inexpensive abundance? We have expanded our numbers and our per-capita consumption rates of virtually all resources. We have created widening waste streams, and we have imperiled the existence of nearly every ecosystem on the planet. Why would more "free" energy lead to anything other than more of the same? Even if we hypothesize a completely nonpolluting energy source, we would still need to eat, and we would still need raw materials of various kinds in order to maintain our still-growing numbers in the way of life to which we have become — or would like to become — accustomed. The rate of species extinctions would continue to escalate, and at some point in the not-too-distant future we would encounter an ecological crisis that threatened the continued existence of the species that matters most to us.

But what, then, is the answer? An analogy may be helpful. Suppose a man wins the lottery and suddenly finds himself in the possession of 10 million dollars. He uses the money to buy a penthouse apartment in Manhattan and a fleet of Italian sports cars; he gambles in Las Vegas; he develops expensive tastes in food, art, and clothing. Then one day he notices that he has only a few hundred dollars left in his bank account. Meanwhile his four children are nearing college age and are pestering him about enrolling in expensive schools. What is he to do? Let's say he imagines that the solution is simply to win the lottery again, and so he begins buying more lottery tickets. In that case, the story is unlikely to have a happy ending. In reality, his best option is to sell the penthouse and cars, buy a modest home, and get a job.

I would suggest that the effort to find more sources of cheap energy is somewhat analogous to buying more lottery tickets. Even if we "win," we will simply be miring ourselves deeper in a fundamentally unsustainable mode of existence.

Thus there may be no solution to the problem of oil depletion, if by "solution" we mean a strategy that will enable us to continue living as we are. "Free" energy has enabled us to create a lifestyle that has no future, simply because it is predicated on unending growth, and continuous growth within a finite system is an impossibility.

This information may be hard to take, but take it we must. There are problems in life that can be solved and those that can't. If the problem is that the register in our checkbook hasn't been kept in order, that is a problem we can solve — though possibly only with considerable effort. If the problem is that we are getting older and cannot do all of the things we could when we were young, we are fighting a losing battle. There are better and worse strategies in that case: we could improve our diet and get more exercise, in which case we would prolong our youthfulness as long as possible. Or we could spend our days smoking cigarettes, eating junk food, and watching hours of television, in which case we would squander and shorten whatever time we had left.

Similarly, with oil depletion there is no solution — in that there is no way to substitute something else for oil and then continue as we are, which means continually growing our population and economy. But there are better and worse ways to respond to the challenge. If we were smart, we could do the equivalent of moving into a modest home and getting a job; we could improve our diets and start getting more exercise. That is, we could begin systematically and cooperatively to reduce our population and per-capita resource consumption, re-localize our economies, and maximize the efficiency of our energy usage. (I offered more specific prescriptions along these lines in Chapter 6.) Better solar panels or wind turbines could help in the transition, but only (and I must stress and re-stress the word *only*) if adopted in the context of a worldwide effort to simplify and downsize the human project.

Meanwhile, the cult of the inventor-savior only mires us deeper in denial. It gives us hope of redemption and of paradise regained — but it is a false and poisonous hope, because it distracts us from taking the intelligent though difficult actions that offer us the best chance of surviving the depletion of fossil fuels.[2]

Where the Real Hope Lies

Many other readers contacted me to say that my book is depressing. I am sorry if this is the case, but that was certainly not the intent. My aim has been simply to alert as many people as possible to a profound change that is about to overtake our civilization and our way of life. In Chapter 6, I did try to offer positive suggestions of things that people can do to help their families, communities, and nations survive the coming energy famine. In the end, optimism is most useful as a state of mind that fosters constructive action. It is self-delusional to dwell on hopeful images of the future merely to distract ourselves from facing unpleasant truths or to avoid having to take difficult actions.

While the international political scene looked bad enough as I was writing the original edition of this book (and, as I have explained above, it looks even more worrisome today), at least the subject of global oil peak is quickly getting out to a larger audience. This increased awareness will not by itself lead us toward a survivable future, but it is an essential prerequisite.

I still believe that if the people of the world can be helped to understand the situation we are in, the options available, and the consequences of the path we are currently on, then it is at least possible that they can be persuaded to undertake the considerable effort and sacrifice that will be entailed in a peaceful transition to a sustainable, locally based, decentralized, low-energy, resource-conserving social regime. But inspired leadership will be required. Everywhere I have traveled to speak on this subject, audiences have shown not just a willingness, but an almost heart-wrenching eagerness to be part of such a collective undertaking. Until inspired leadership does emerge, we must do what we can at the local level, wherever we are.

Notes

Introduction

1. Robert M. Solow, quoted in Herman Daly, *Steady-State Economics* (Island Press, 1991), p. 117.

1 — Energy, Nature, and Society

1. "In 1995, researchers found bacteria subsisting on rock and water about 1,000 meters (3,200 feet) down in aquifers within volcanic rocks near the Columbia River in Washington. They survive by getting dissolved CO_2 from the groundwater and appear to get energy by using hydrogen (H_2) generated in a reaction between iron-rich minerals in the rock and groundwater." G. Tyler Miller, Jr., *Living in the Environment: Principles, Connections, and Solutions* (Brooks/Cole, 2002), p. 100.

2. However, it might be argued that, due to our present reliance on fossil fuels for agricultural food production, we humans have also become, in a sense, scavengers or detritovores.

3. For examples of avoidance of competition in nature, see Robert Augros and George Stanciu, *The New Biology: Discovering the Wisdom in Nature* (Shambhala, 1987), pp. 91-105.

4. Quoted in Augros and Stanciu, *The New Biology*, p. 118.

5. Lynn Margulis and James Lovelock, "Is Mars a Spaceship, Too?" *Natural History*, June/July 1976, pp. 86-90.

6. These strategies are described at length by sociologist William Catton in his groundbreaking book *Overshoot: the Ecological Basis of Revolutionary Change* (University of Illinois Press, 1980).

7. William Catton, *Overshoot*, p. 27.

8. For a more detailed and highly readable description of this process, see Marvin Harris, *Cannibals and Kings: The Origins of Cultures*, (Random House, 1977), pp. 9-30.

9. Theodore Xenophon Barber, *The Human Nature of Birds* (St. Martin's Press, 1993), p. 12

275

10. Associated Press, 8 January 1999.

11. Quoted in David Suzuki and Peter Knudtson, *Wisdom of the Elders: Sacred Native Stories of Nature* (Bantam, 1992), p. 212.

12. Theodore Xenophon Barber, *The Human Nature of Birds*, p. 11.

13. While this idea has been expressed elsewhere, William Catton developed it brilliantly in *Overshoot* (pp. 143-155), and I follow his exposition closely.

14. William Catton, *Overshoot*, p. 150.

15. Ibid., p. 146.

16. See Marshall McLuhan, *The Gutenberg Galaxy* (Routledge & Kegan Paul, 1962); see also, for example, Plato, *Phaedrus*, verses 67-71.

17. William Catton, *Overshoot*, p. 159.

18. John H. Lienhard, *Power Production,* [online], <www.uh.edu/engines/epi277.htm>

19. Julie Wakefield, "Boys Won't Be Boys," *New Scientist,* 29 June 2002, pp. 42-5.

20. Ross Gelbspan, *The Heat Is On: The Climate Crisis, the Coverup, the Prescription* (Perseus, 1998).

21. Joseph Tainter, "Complexity, Problem Solving, and Sustainable Societies," in Robert Costanza et al., eds., *Getting Down to Earth: Practical Applications of Ecological Economics* [online], (Island Press, 1996), <http://dieoff.com/page134.htm>

22. Joseph Tainter, "Complexity, Problem Solving, and Sustainable Societies".

23. Joseph Tainter, *The Collapse of Complex Societies* (Cambridge University Press, 1988), pp. 91-2.

24. Joseph Tainter, *The Collapse of Complex Societies*, p. 195.

25. Ibid., p. 196.

26. Joseph Tainter, "Complexity, Problem Solving, and Sustainable Societies".

27. Joseph Tainter, *The Collapse of Complex Societies*, p. 216.

28. Jared Diamond, *Guns, Germs, and Steel: The Fates of Human Societies* (W. W. Norton, 1999).

29. William H. McNeill, *Plagues and Peoples* (Anchor, 1998), pp. 31-68, 176-207.

30. Jared Diamond, *Guns, Germs, and Steel*, p. 373.

2 — Party Time: The Historic Interval of Cheap, Abundant Energy

1. Marvin Harris, *Cannibals and Kings: The Origins of Cultures* (Random House, 1977), p. 25; see also Jared Diamond, *Guns, Germs, and Steel*.

2. Quoted in Lewis Mumford, *The Myth of the Machine, Vol. I: Technics and Human Development* (Harcourt Brace Jovanovich, 1966), p. 247.

3. Quoted in Fernand Braudel, *The Structures of Everyday Life: The Limits of the Possible* (University of California Press, 1992), p. 366.

4. Fernand Braudel, *The Structures of Everyday Life*, p. 368.

5. Quoted in Jeremy Rifkin, with Ted Howard, *Entropy: A New World View* (Bantam, 1981), p. 74.

6. Lewis Mumford, *The Myth of the Machine, Vol. II: The Pentagon of Power* (Harcourt Brace Jovanovich, 1964), p. 147.

7. Quoted in Margaret Cheney and Robert Uth, *Tesla: Master of Lightning* (Friedman/Fairfax, 1999), p. 19.

8. Vaclav Smil, *Enriching the Earth: Fritz Haber, Carl Bosch, and the Transformation of World Food Production* (MIT Press, 2001).

9. Quoted in Katie Alvord, *Divorce Your Car!: Ending the Love Affair with the Automobile* (New Society, 2000), p. 18.

10. See Peter Schweizer, *Victory: The Reagan Administration's Secret Strategy that Hastened the Collapse of the Soviet Union* (Atlantic Monthly, 1996).

11. Jimmy Carter, *Address to the Nation*, 18 April 1977.

12. Helga Graham, "Exposed: Washington's Role in Saddam's Oil Plot," *London Observer*, 21 October 1990; see also Phyllis Bennis and Michel Moushabeck, eds., *Beyond the Storm: A Gulf Crisis Reader* (Olive Branch Press, 1991), p. 395.

13. Stephen C. Pelletiere, *Iraq and the International Oil System: Why America Went to War in the Gulf* (Praeger, 2001).

14. See Jim Vallette and Daphne Wysham, *Enron's Pawns: How Public Institutions Bankrolled Enron's Globalization Game* [online], Institute for Policy Studies Report, March 2002, <www.seen.org/pages/reports.shtml>.

15. However, economic inequality was greatly tempered by government intervention in the economy. In the US, the progressive income tax and various government programs stemming from the New Deal reduced economic inequality substantially during the middle decades of the century. But then the reduction of tax rates for the wealthy and the gradual undermining of social programs by the Reagan-Bush administration in the 1980s signaled the beginning of a steep rise in inequality that continues up to the present.

16. See Richard Douthwaite, "When Should We Have Stopped?", *Irish Times* [online], 29 December 2001,
    <www.greenbooks.co.uk/Douthwaite/machines.htm>.

17. Widely quoted on 30 April 2001.

3 — Lights Out: Approaching the Historic Interval's End

1. Throughout this book the word *terrorism* appears within quotation marks because I wish to call attention to the fact that it is a highly politicized term with no uniform definition. A given act —such as an assassination or the bombing of civilians — is officially declared by the US government to be "terrorism" when committed by one organized group, whereas a similar act committed by another group is labeled "counterterrorism" or "self-defense." Since the word so dominates public discourse, it is scarcely possible to avoid using it; the implied irony of quotation marks is simply a reminder that the term always carries an unstated political agenda and point of view.

2. Jean-Charles Brisard and Guillaume Dasquie, *Forbidden Truth: U.S.-Taliban Secret Oil Diplomacy, Saudi Arabia, and the Failed Search for bin Laden* (Thunder's Mouth, 2002).

3. Jean-Charles Brisard and Guillaume Dasquie, *Forbidden Truth;* see also *Greg Palast, The Best Democracy Money Can Buy: An Investigative Reporter Exposes the Truth about Globalization, Corporate Cons, and High Finance Fraudsters* (Pluto, 2002).

4. Goldman Sachs, *Energy Weekly,* 11 August 1999.

5. EIA Annual Energy Outlook 2000 with Projections to 2020; current report available online at <www.eia.doe.gov/oiaf/aeo/>.

6. Blair made these comments on January 28, 2000, at Davos, Switzerland. See, for example, <http://Singapore.emc/com/news/in_depth_archive/01032001_year_info.jsp>.

7. Eric Haseltine, "Twenty Things That Will Be Obsolete in Twenty Years," *Discover,* Vol. 21, No. 10, October 2000, p. 85.

8. William G. Phillips, "Are We Really Running Out of Oil?", *Popular Science,* May 2000, p. 56.

9. Hubbert was fortunate to deal with the US lower-48, where a simple bell-shaped curve could be derived from production statistics. Adding production data from Alaska changes the curve, revealing two cycles. Further, between 1960 and 1980 supply was almost never constrained by demand — because of mandatory quotas on oil imports instituted by Eisenhower in 1959. Without these quotas, US production would have been less since foreign oil was cheaper. Thus reduced demand for more expensive domestic oil would have delayed the peak.

10. Kenneth S. Deffeyes, *Hubbert's Peak: The Impending World Oil Shortage* (Princeton University Press, 2001), pp. 134-49.

11. Kenneth S. Deffeyes, *Hubbert's Peak,* p. 135

12. "Two Intellectual Systems: Matter-energy and the Monetary Culture." Summary, by M. King Hubbert, of a seminar he taught at MIT Energy Laboratory, 30 September 1981, at

<www.hubbertpeak.com/hubbert/monetary.htm>.

13. See <www.hubbertpeak.com/hubbert/hubecon.htm>.

14. See <www.hubbertpeak.com/hubecon.htm>.

15. See <www.technocracy.org/articles/hubbert-econ.html>.

16. C. J. Campbell, *The Coming Oil Crisis* (Multi-Science Publishing Company and Petroconsultants, 1997).

17. C. J. Campbell and Jean Laherrère, "The End of Cheap Oil?", *Scientific American* [online], March 1998, <http://dieoff.org/page140.htm>.

18. C. J. Campbell, *Peak Oil: An Outlook on Crude Oil Depletion* [online], October 2000, <www.mbendi.co.za/indy/oilg/p0005.htm>.

19. Kenneth S. Deffeyes, *Hubbert's Peak*, p. 149.

20. L. F. Ivanhoe, "King Hubbert Updated," *Hubbert Center Newsletter* [online], No. 97/1, <http://hubbert.mines.edu/>.

21. Walter Youngquist, *Geodestinies: The Inevitable Control of Earth Resources over Nations and Individuals* (National, 1997), p.183.

22. Walter Youngquist, *Geodestinies*, p. 200.

23. See <www.simmonsco-intl.com>.

24. From a personal communication with the author.

25. See <www.bp.com/subsection.do?categoryId=95&contentId=2006480>.

26. From a personal communication with author.

27. See <www.odac-info.org>.

28. Peter Huber, "The Energy Spiral," Forbes [online], 1 April 2002, <www.forbes.com/forbes/2002/0401/102.htm>.

29. Bjørn Lomborg, "Running on Empty," *The Guardian* [online], 16 August 2001, <www.guardian.co.uk/Archive/Article/0,4273,4239923,00.html>.

30. The previous two sentences are adapted from remarks by Tom Robertson, Moderator of the online EnergyResources news group, 15 March 2002.

31. Walter Youngquist, *Geodestinies*, p. 222.

32. Ibid., p. 216.

33. For more details about this process, see <http://dieoff.com/page143.htm>.

34. Craig Campbell and Emilia Kennedy, "Alberta Oil Extraction Hurts Native People," *Earth Island Journal* [online], Vol. 17, No. 3, Autumn

2002, p. 6, <www.earthisland.org/eijournal/new_articles.cfm?articleID= 596&journalID=65>.

35. See <http://sepwww.stanford.edu/sep/jon/world-oil.dir/lynch2.html>.

36. USGS website: <www.usgs.gov>; USGS World Petroleum Assessment 2000: <http://greenwood.cr.usgs.gov/pub/fact-sheets/fs-0070-00/>. For a more detailed analysis of the USGS report, see <www.oilcrisis.com/ duncan/usgs2000.htm>.

37. Energy Information Agency, *Annual Energy Outlook 1998 with Projections to 2020*, p. 17.

38. See < www.iea.org/weo/insights.htm>.

39. International Energy Agency, *World Energy Outlook: 2001* [online], <www.iea.org/weo/insights.htm>.

4 — Non-Petroleum Energy Sources: Can the Party Continue?

1. Howard T. Odum, *Environmental Accounting, Emergy and Decision Making* (John Wiley, 1996); C. J. Cleveland, R. Costanza, C. A. S. Hall, and R. Kaufmann, "Energy and the U.S. Economy: A Biophysical Perspective," Science 225 1984, 890-97, rpt. in John Gever, Robert Kaufman, David Skole, and Charles Vorosmarty, *Beyond Oil: The Threat to Food and Fuel in the Coming Decades* (University Press of Colorado, 1991), p. 70.

2. See <www.eia.doe.gov/emeu/iea/table81.html>.

3. Brad Foss, *Companies struggle to keep steady production of natural gas* [online], <www.detnews.com/2001/business/0108/10/b03-265559.htm>.

4. Gary S. Swindell, "Texas Production Data Show Rapid Gas Depletion," *Oil & Gas Journal* [online], 21 June 1999, p. 61, <http://oil.server4.com/tx-depl.htm>.

5. Randy Udall and Steve Andrews, *Natural Gas Primer* [online], <www.oilcrisis.com/gas/primer/>.

6. Julian Darley, *High Noon for Natural Gas: The New Energy Crisis* (Chelsea Green Publishing, 2003), p. 2.

7. John Gever *et al.*, *Beyond Oil*, pp. 65-8.

8. See <www.copvcia.com/free/ww3/052504_coal_peak.html>.

9. Walter Youngquist, *Geodestinies*, p. 224.

10. See <www.sortirdunucleaire.org>.

11. Doug Dupler, *Energy: Shortage, Glut, or Enough?* (Information Plus, 2001), p. 65.

12. See P. H. Eichstaedt, *If You Poison Us: Uranium and Native Americans* (Crane, 1994); see also <www.cpluhna.nau.edu/Change/uranium.htm>.

13. CNN, 8 May 2001; see also <www.nirs.org/factsheets/KYOTONUC.html>.

14. *International Herald Tribune,* 8 May 2001; see also <www.antenna.nl/~wise/uranium/nfp.html>.

15. See <www.andra.fr> and <www.nirs.org>.

16. See <www.nei.org>.

17. *Nucleonics Week* [online], 10 May 2001, <www.pbmr.com>.

18. See <www.wind-sail.com>.

19. See <www.nrel.gov/wind/potential.html>. The study assumed a 25-percent turbine efficiency, with 25 percent losses, and made exclusions for environmental, urban, and agricultural purposes.

20. See <www.regenesys.com/ourproducts/index.htm>.

21. See <www.EON_Netz_Windreport_e_eng_.pdf>.

22. See <www.windpower.dk/tour/env/enpaybk.htm>.

23. Spheral: <www.psheralsolar.com>; CIGS: <www.nrel.gov/ncpv/hotline/01_01_solar_cell.html>.

24. See <www.isr.gov.au/resources/netenergy/aen/aen20/10titania.html>.

25. See <www.futurepundit.com/archives/002230.html>.

26. *Applied Physics Letters,* November 29, 2004.

27. See <www.Ecotopia.com/apollo2/pvepbtne.htm>.

28. David Stipp, "The Coming Hydrogen Economy," *Fortune* [online], 12 November 2001, <www.business2.com/articles/mag/print/0,1643,34966,00.html>.

29. Quoted in *Earth Island Journal* [online], Vol. 6, No. 2, Summer 2001, <www.earthisland.org/eijournal/new_articles.cfm?articleD=137&journalID=46>.

30. See <www.metallicpower.com>.

31. See <www.hypercar.com/index.html> and <www.mercedes-benz.com/e/innovation/fmobil/necar3.htm>.

32. See <www.rmi.org/sitepages/pid311.php>.

33. See <www.microhydropower.net>.

34. See <www.bluenergy.com>.

35. David Ross, *Power from the Waves* (Oxford University Press, 1995). See also Walter Youngquist, *Geodestinies,* p. 253; and Howard T. Odum and

Elizabeth C. Odum, *A Prosperous Way Down: Principles and Policies* (University Press of Colorado, 2001), p. 162.

36. Pimentel, David, "Ethanol Fuels: Energy Balance, Economics, and Environmental Impacts Are Negative," *Natural Resources Research*, Vol. 12, No. 2, June 2003.

37. See <www.cornandsoybeandigest.com/news/EthanolValue/>.

38. Tad W. Patzel, "Critical Reviews in Plant Sciences," University of California, Berkeley, October 30, 2004.

39. Scott Kilman, "US: Surging Imports of Food Threaten Wider Trade Gap," *Wall Street Journal*, November 8, 2004.

40. See <www.infinite-energy.com>.

41. John Gever et al., *Beyond Oil*, p. 101-2.

42. Ibid., pp. 102-3.

43. Three excellent books on this topic are Bill Devall, *Living Richly in an Age of Limits* (Gibbs-Smith, 1993); Joe Dominguez and Vicki Robin, *Your Money or Your Life: Transforming Your Relationship With Money and Achieving Financial Independence* (Penguin, 1999); and Paul Wachtel, *Poverty of Affluence: A Psychological Study of the American Way of Life* (Free Press, 1983).

44. Howard T. Odum and Elisabeth C. Odum, *A Prosperous Way Down*, p. 169.

5 — A Banquet of Consequences

1. Katie Alvord, *Divorce Your Car*, p. 165.

2. Lester Brown, *Who Will Feed China?: Wake-up Call for a Small Planet* (W. W. Norton, 1995), p. 123.

3. Nicholas Parrott and Terry Marsden, *The Real Green Revolution: Organic and Agroecological Farming in the South* [online], Greenpeace, 2002, p. 62, <www.greenpeace.de/GP_DOK_3P/BRENNPUN/F0107N11.PDF>.

4. Monsanto website: <www.monsanto.com/monsanto/media/speeches/new_pledge_speech.html>, accessed June 7, 2002.

5. Based on John Jeavons' statements made during a tour of the Ecology Action demonstration site.

6. Agricultural biotechnology relies upon industrial farming methods. Reductions in pesticide and herbicide applications resulting from the use of genetically engineered seeds are only a minor benefit in terms of fossil-fuel savings and could be duplicated through agro-ecological methods.

7. Chris Adams, *Heat and Cold: The Next Step in Effective Warning* [online], <www.colorado.edu/hazards/ss/ss97/5.htm>.

8. All figures are from David Pimentel et al., "Ecology of Increasing Disease: Population growth and environmental degradation," *Bioscience*, Vol. 48, No. 10, October 1998, pp. 817-27.

9. Based on data and projections by J. Lederberg, R. E. Shope, and S. C. Oaks, *Emerging Infections: Microbial Threats to Health in the United States* (National Academy Press, 1992); see also World Health Organization, *World Health Report*, 1996.

10. Mark Schoofs, "AIDS: The Agony of Africa," *The Village Voice* [online], 3-9 November 1999, <www.villagevoice.com/specials/africa/>.

11. See Laurie Garrett and Steven Wolinsky, *Betrayal of Trust: The Collapse of Global Public Health* (Hyperion, 2001).

12. See Paul H. Ray and Sherry Ruth Anderson, *The Cultural Creatives: How 50 Million People Are Changing the World* (Three Rivers Press, 2001).

13. This analysis is most thoroughly developed in the writings of researchers investigating population/resource dynamics, such as Paul R. and Anne H. Ehrlich, David Pimentel, and Garrett Hardin (see Bibliography).

14. See John C. Stauber and Sheldon Rampton, *Toxic Sludge Is Good for You: Lies, Damn Lies, and the Public Relations Industry* (Common Courage, 1995).

15. See Lawrence H. Keeley, *War Before Civilization* (Oxford University Press, 1996) and Raymond C. Kelly, *Warless Societies and the Origin of War* (University of Michigan Press, 2000).

16. The National Endowment for Democracy — a private nonprofit organization that is a front for the State Department and has already been implicated numerous times in the provision of money to sway elections (in Chile in 1988, Nicaragua in 1989, and Yugoslavia in 2000) — reportedly gave $877,000 during 2001 to forces opposed to Chavez. More than $150,000 went to Carlos Ortega, leader of the corrupt Confederation of Venezuelan Workers, who worked closely with the coup's leader, Pedro Carmona Estanga. Bush administration officials had met in Washington with anti-Chavez Venezuelan generals and businessmen in the weeks preceding the coup, and George W. Bush's Assistant Secretary of State for Western Hemisphere Affairs, Otto Reich, was reported to have been in contact with the civilian head of the junta on the day of the coup. When Venezuelans took to the streets to defend their popular president and he was restored to power, US officials grudgingly acknowledged that Chavez had been freely elected (with 62 percent of the vote) although one told a reporter that "legitimacy

is something that is conferred not just by a majority of the voters" — a remark that held a peculiarly ironic resonance given the process and results of the most recent US presidential election.

17. Julia Olmstead, *In Colombia, US Companies Get Down to Business* [online], 5 July 2002, <www.americas.org/>.

18. Michael Klare, *Resource Wars: The New Landscape of Global Conflict* (Metropolitan, 2001), pp. 112-3.

19. Blair is said to have told close friends that, in a meeting between himself, Bush, and French President Jacques Chirac, the subject of France's economy came under discussion; at one point, Bush leaned over to Blair and whispered, in all seriousness, "The problem with the French is that they don't have a word in their language for *entrepreneur.*"

    See <www.j-bradford-delong.net/movable_type/archives/000349.html> dated 14 July 2002.

20. Paul Stuart, *Camp Bondsteel and America's plans to control Caspian oil* [online], 29 April 2002, <www.wsws.org/articles/2002/apr2002/oil-a29.shtml>.

21. See Noam Chomsky, *The Umbrella of U.S. Power: The Universal Declaration of Human Rights and the Contradictions of U.S. Policy* (Seven Stories Press, 1999).

22. Indra de Soysa, quoted in "Blood, Diamonds and Oil," (no author listed), *New Scientist,* 29 June 2002, p. 36.

23. Richard C. Duncan, *The Peak of World Oil Production and the Road to the Olduvai Gorge* [online], 2000, <www.hubbertpeak.com/duncan/olduvai2000.htm>.

24. Ernest Partridge, *Perilous Optimism* [online], <http://gadfly.igc.org/papers/cornuc.htm>.

25. Leslie A. White, *The Science of Culture: A Study of Man and Civilization* (Farrar, Straus & Co., 1949).

26. Richard Duncan, *The Peak of World Oil Production and the Road to Olduvai Gorge.*

27. This observation was first noted and brilliantly developed by Richard Duncan in his book *The Peak of World Oil Production and the Road to Olduvai Gorge.*

6 — Managing the Collapse: Strategies and Recommendations

1. Donella H. Meadows, Dennis L. Meadows, and Jørgen Randers, *Beyond the Limits: Confronting Global Collapse, Envisioning a Sustainable Future* (Chelsea Green, 2002), p. 136.

2. Donella H. Meadows *et al.*, *Beyond the Limits*, p. xvi.

3. Ibid., p. 222.

4. Hartmut Bossel, *Earth at a Crossroads: Paths to a Sustainable Future* (Cambridge University Press, 1998), p. 8.

5. Howard T. Odum and Elizabeth C. Odum, *A Prosperous Way Down*, pp. 3-5.

6. Joe Jenkins, *Humanure Handbook* (Jenkins Publishing, 1999).

7. See Mathis Wackernagel and William Rees, *Our Ecological Footprint: Reducing Human Impact on the Earth* (New Society, 1996).

8. For information see <www.endofsuburbia.com>.

9. For a more detailed discussion of the Bush energy plan, see <www.azsolarcenter.com/calendar/expanded/sa-5-25-01-559.html>.

10. Frida Berrigan,"Sky High: The military busts the 2003 federal budget," *In These Times* [online], March 2002, <www.thirdworldtraveler.com/Pentagon_military/SkyHigh_2003MilitaryBudget.html>.

11. Michael Pollan, "When a Crop Becomes King," *New York Times,* 19 July 2002.

12. Philipp Lee, *Killing the land with farm subsidies* [online], 19 July 2002, <www.fcpp.org/publications/fcpp_media/farm_subsidies.html>.

13. Joel Bleifuss, "Know Thine Enemy: A Brief History of Corporations," *In these Times* [online], February 1998, <www.thirdworldtraveler.com/Corporations/KnowEnemy_ITT.html>.

14. Hans Krampe, *Bunks: The Fox In Charge of the Perfect Golden Goose, or The Debt-Based Banking System* [online], <www.radicalpress.com/writers/hk_article01.html>.

15. Russell Hopfenberg and David Pimentel, "Human Population Numbers as a Function of Food Supply," *Environment, Development and Sustainability* [online], Vol. 3 (Kluwer, 2001), <http://216.239.53.100/search?q=cache:DH3rwQ9RQdcC:www.ku.edu/~hazards/foodpop.pdf+pimentel+hopfenberg+human+population&hl=en&ie=UTF-8>.

16. Ibid.

17. Garrett Hardin, *Living Within Limits: Ecology, Economics, and Population Taboos* (Oxford University Press, 1993), p. 269.

18. Garrett Hardin, *Living Within Limits,* p. 271.

19. Ibid., p. 272

20. Ibid., p. 279

21. Virginia Abernethy, *The Demographic Transition Revisited: Lessons for Foreign Aid and U.S. Immigration Policy* [online], <www.carryingcapacity.org/va2.html>.

22. George Kennan, US State Department Policy Planning Study #23, (1948), quoted in John Pilger, *Hidden Agendas* (The New Press, 1998), p.59.

23. Jimmy Carter, *State of the Union Address*, January 1980.

24. William Blum, *Rogue State: A Guide to the World's Only Superpower* (Common Courage, 2002).

25. Ohio Association of Rail Passengers, *America's long history of subsidizing transportation* [online], <www.trainweb.org/moksrail/advocacy/resources/subsidies/transport.htm>.

26. Garrett Hardin, *Living Within Limits*, p. 278.

27. See Derrick Jensen, *The Culture of Make Believe* (Context, 2002).

Afterword to the Revised Edition

1. See <www.peakoil.net/Channel4.html>.

2. I owe Matt Savinar of <www.lifeaftertheoilcrash.net> a debt of gratitude for some of the ideas in this section, which emerged in conversation over lunch.

Bibliography

Abernethy, Virginia. *The Demographic Transition Revisited: Lessons for Foreign Aid and U.S. Immigration Policy* [online]. www.carryingcapacity.org/va2.html>.

Abi-Nader, Jeanette, Kendall Dunnigan, and Kristen Markley. *Growing Communities Curriculum: Community Building and Organizational Development through Community Gardening.* The American Community Gardening Association, 2001.

Adams, Chris. *Heat and Cold: The Next Step in Effective Warning* [online]. <www.colorado.edu/hazards/ss/ss97/5.htm>.

Alvord, Katie. *Divorce Your Car!: Ending the Love Affair with the Automobile.* New Society, 2000.

Augros, Robert, and George Stanciu. *The New Biology: Discovering the Wisdom in Nature.* Shambhala, 1987.

Barber, Theodore Xenophon. *The Human Nature of Birds.* St. Martin's Press, 1993.

Bennis, Phyllis, and Michel Moushabeck, eds. *Beyond the Storm: A Gulf Crisis Reader.* Olive Branch Press, 1991.

Berman, Daniel M., and John O'Connor. *Who Owns the Sun? People, Politics and the Struggle for a Solar Economy.* Chelsea Green, 1996.

Berrigan, Frieda. "Sky High: The military busts the 2003 federal budget." *In These Times* [online], March 2002. <www.thirdworldtraveler.com/ Pentagon_military/SkyHigh_2003MilitaryBudget.html>.

Bleifuss, Joel. "Know Thine Enemy: A Brief History of Corporations." *In these Times* [online], February 1998. <www.thirdworldtraveler.com/ Corporations/KnowEnemy_ITT.html>.

Blum, William. *Rogue State: A Guide to the World's Only Superpower.* Common Courage, 2000.

Bossel, Hartmut. *Earth at a Crossroads: Paths to a Sustainable Future.* Cambridge University Press, 1998.

Braudel, Fernand. *The Structures of Everyday Life: The Limits of the Possible.* University of California Press, 1992.

Brisard, Jean-Charles, and Guillaume Dasquie. *Forbidden Truth: U.S.-Taliban Secret Oil Diplomacy, Saudi Arabia, and the Failed Search for bin Laden.* Thunder's Mouth, 2002.

Brown, Lester. *Who Will Feed China?: Wake-up Call for a Small Planet.* W. W. Norton, 1995.

Brzezinski, Zbigniew. T*he Grand Chessboard: American Primacy and Its Geostrategic Imperatives* . Basic Books, 1997.

Burch, Mark *A. Stepping Lightly: Simplicity for People and the Planet.* New Society, 2000.

Burke, James, and Robert Ornstein. *The Axemaker's Gift: Technology's Capture and Control of Our Minds and Culture.* Tarcher/Putnam, 1995.

Campbell, C. J. *The Coming Oil Crisis.* Multi-Science and Petroconsultants, 1997.

Campbell, C.J. "Running Out of Gas: This Time the Wolf Is Coming." *The National Interest,* No. 51, Spring 1998, pp. 47-55.

Campbell, C.J. *The Future of Oil and Hydrocarbon Man.* Multi-Science and Petroconsultants, n.d.

Campbell, C.J. Peak Oil: *An Outlook on Crude Oil Depletion* [online]. October 2000. <www.mbendi.co.za/indy/oilg/p0005.htm>.

Campbell, C.J. and Jean Laherrère. "The End of Cheap Oil?", *Scientific American* [online], March 1998. <http://dieoff.org/page140.htm>.

Campbell, Craig, and Emilia Kennedy. "Alberta Oil Extraction Hurts Native People." *Earth Island Journal* [online], Vol. 17, No. 3, Autumn 2002. <www.earthisland.org/eijournal/new_articles.cfm?articleID= 596&journalID=65>.

Catton, William. *Overshoot: The Ecological Basis of Revolutionary Change.* University of Illinois Press, 1980.

Cheney, Margaret, and Robert Uth. *Tesla: Master of Lightning.* Friedman/ Fairfax, 1999.

Chomsky, Noam. Rogue States: *The Rule of Force in World Affairs.* South End Press, 2000.

Chomsky, Noam. *The Umbrella of U.S. Power: The Universal Declaration of Human Rights and the Contradictions of U.S. Policy.* Seven Stories Press, 1999.

Daly, Herman. *Steady-State Economics.* Island Press, 1991.

Deffeyes, Kenneth S. *Hubbert's Peak: The Impending World Oil Shortage.* Princeton University Press, 2001.

Devall, Bill. *Living Richly in an Age of Limits.* Gibbs-Smith, 1993.

Diamond, Jared. *Guns, Germs, and Steel: The Fates of Human Societies.* W. W. Norton, 1999.

Dominguez, Joe, and Vicki Robin. *Your Money or Your Life: Transforming Your Relationship with Money and Achieving Financial Independence.* Penguin, 1999.

Douthwaite, Richard. "When Should We Have Stopped?" Irish Times [online], 29 December 2001. <www.greenbooks.co.uk/Douthwaite/machines.htm>.

Douthwaite, Richard. *The Growth Illusion: How Economic Growth Enriched the Few, Impoverished the Many, and Endangered the Planet.* New Society, 1992.

Duncan, Richard C. *The Peak of World Oil Production and the Road to the Olduvai Gorge* [online]. <www.hubbertpeak.com/duncan/olduvai2000.htm>.

Dupler, Doug. Energy: *Shortage, Glut, or Enough?* Information Plus, 2001.

Ehrlich, Paul R. and Anne H. *End of Affluence.* Ballantine, 1980.

Eichstaedt, P. H. *If You Poison Us: Uranium and Native Americans.* Crane, 1994.

Fleming, David. *After Oil* [online]. <www.geocities.com/davidmdelaney/after-oil-david-fleming.html>.

Foss, Brad. *Companies struggle to keep steady production of natural gas* [online]. <www.detnews.com/2001/business/0108/10/b03-265559.htm>.

Garrett, Laurie, and Steven Wolinsky. *Betrayal of Trust: The Collapse of Global Public Health.* Hyperion, 2001.

Gastil, John. *Democracy in Small Groups: Participation, Decision-making and Communication.* New Society, 1993.

Gelbspan, Ross. *The Heat Is On: The Climate Crisis, the Coverup, the Prescription.* Perseus, 1998.

Gever, John, Robert Kaufmann, David Skole, and Charles Vorosmarty. *Beyond Oil: The Threat to Food and Fuel in the Coming Decades.* University Press of Colorado, 1991.

Gladstar, Rosemary. *Herbs for the Home Medicine Chest.* Storey, 1999.

Gliessman, Stephen R. *Agroecology: Ecological Processes in Sustainable Agriculture.* Lewis, 1998.

Goering, Peter, Helena Norberg-Hodge, and John Page. *From the Ground Up: Rethinking Industrial Agriculture.* 2nd ed. ISEC, 2000.

Graham, Helga. "Exposed: Washington's Role in Saddam's Oil Plot." *London Observer*, 21 October 1990.

Greco, Thomas H., Jr. *New Money for Healthy Communities.* Greco, 1994.

Greco, Thomas H. *Understanding and Creating Alternatives to Legal Tender.* Chelsea Green, 2002.

Hardin, Garrett. *Living Within Limits: Ecology, Economics, and Population Taboos.* Oxford University Press, 1993.

Harris, Marvin. *Cannibals and Kings: The Origins of Cultures.* Random House, 1977.

Hartmann, Thom. *The Last Hours of Ancient Sunlight: Waking Up to Personal and Global Transformation.* Mythical Books, 1998.

Haseltine, Eric. "Twenty Things That Will Be Obsolete in Twenty Years." *Discover,* Vol. 21, No. 10, October 2000, p. 84-7.

Hawken, Paul, Amory Lovins, and L. Hunter Lovins. *Natural Capitalism: Creating the Next Industrial Revolution.* Back Bay, 2000.

Hopfenberg, Russell, and David Pimentel. "Human Population Numbers as a Function of Food Supply." *Environment, Development and Sustainability* [online]. Vol. 3, Kluwer, 2001. <http://216.239.53.100/search?q= cache:DH3rwQ9RQdcC:www.ku.edu/~hazards/foodpop.pdf+pimentel+ hopfenberg+human+population&hl=en&ie=UTF-8>.

Hubbert, M. King. *Resources and Man.* National Academy of Sciences and National Research Council, 1969.

Huber, Peter. *Hard Green: Saving the Environment from the Environmentalists.* Basic Books, 1999.

Huber, Peter. "The Energy Spiral," Forbes [online], 1 April 2002. <www.forbes.com/forbes/2002/0401/102.htm>.

International Energy Agency. *World Energy Outlook:* 2001[online]. <www.iea.org/weo/insights.htm>.

Ivanhoe, L. F. "King Hubbert — Updated." *Hubbert Center Newsletter* [online], No. 97/1. <http://hubbert.mines.edu/>.

Jeavons, John, and Carol Cox. *The Sustainable Vegetable Garden: A Backyard Guide to Healthy Soil and Higher Yields.* Ten Speed Press, 1999.

Jenkins, Joe. *Humanure Handbook.* Jenkins Publishing, 1999.

Jensen, Derek. *The Culture of Make Believe.* Context, 2002.

Keeley, Lawrence H. *War Before Civilization.* Oxford University Press, 1996.

Kelly, Raymond C. *Warless Societies and the Origin of War.* University of Michigan Press, 2000.

Kennedy, Joseph E., Michael G. Smith, and Catherine Waneck, eds. *The Art of Natural Building: Design, Construction, Resources.* New Society, 2002.

Klare, Michael. *Resource Wars: The New Landscape of Global Conflict.* Metropolitan, 2001.

Krampe, Hans. *Bunks: The Fox In Charge of the Perfect Golden Goose, or The Debt-Based Banking System* [online]. <www.radicalpress.com/writers/hk_article01.html>.

Lederberg, J., R. E. Shope, and S. C. Oaks. *Emerging Infections: Microbial Threats to Health in the United States.* National Academy Press, 1992.

Lee, Philipp. *Killing the land with farm subsidies* [online], 19 July 2002. <www.fcpp.org/publications/fcpp_media/farm_subsidies.html>.

Lienhard, John H. *Power Production* [online]. <www.uh.edu/engines/epi277.htm>.

Lomborg, Bjørn. *The Skeptical Environmentalist.* Cambridge University Press, 2001.

Lomborg, Bjørn. "Running on Empty." *The Guardian* [online], 16 August 2001. <www.guardian.co.uk/Archive/Article/0,4273,4239923,00.html>.

Lovins, Amory B. *Soft Energy Paths: Toward a Durable Peace.* Harper and Row, 1977.

Margulis, Lynn, and James Lovelock. "Is Mars a Spaceship, Too?" *Natural History*, June/July 1976, pp. 86-90.

Mars, Brigitte. *Natural First Aid: Herbal Treatments for Ailments and Injuries, Emergency Preparedness, Wilderness.* Safety Storey, 1999.

McLuhan, Marshall. *The Gutenberg Galaxy.* Routledge and Kegan Paul, 1962.

McNeill, William H. *Plagues and Peoples.* Anchor, 1998.

Meadows, Donella H., Dennis L. Meadows, and Jørgen Randers, *Beyond the Limits: Confronting Global Collapse, Envisioning a Sustainable Future.* Chelsea Green, 2002.

Merchant, Carolyn. *The Death of Nature: Women, Ecology and the Scientific Revolution.* Harper and Row, 1983.

Miller, G. Tyler, Jr. *Living in the Environment: Principles, Connections, and Solutions.* Brooks/Cole, 2002.

Moyer, Bill, with JoAnn McAllister, Mary Lou Finley, and Steve Soifer. *Doing Democracy: The MAP Model for Organizing Social Movements.* New Society, 2000.

Mumford, Lewis. *The Myth of the Machine, Vol. I: Technics and Human Development.* Harcourt Brace Jovanovich, 1966.

Mumford, Lewis. *The Myth of the Machine, Vol. II: The Pentagon of Power.* Harcourt Brace Jovanovich, 1964.

Nattrass, Brian, and Mary Altomare. *The Natural Step for Business: Wealth, Ecology and the Evolutionary Corporation*. New Society, 2001.

Norberg-Hodge, Helena, Todd Merrifield, and Steven Gorelick. *Bringing the Food Economy Home: Local Alternatives to Global Agribusiness*. Zed, 2002.

Odum, Howard T. and Elisabeth C. *A Prosperous Way Down: Principles and Policies*. University Press of Colorado, 2001.

Odum, Howard T. and Elisabeth C. Odum. *Energy Basis for Man and Nature*. McGraw-Hill, 1976.

Odum, Howard T. *Environment, Power and Society*. Wiley, 1971.

Odum, Howard T. *Environmental Accounting, Emergy and Decision Making*. Wiley, 1996.

Ohio Association of Rail Passengers. *America's long history of subsidizing transportation* [online]. <www.trainweb.org/moksrail/advocacy/resources/subsidies/transport.htm>.

Olmstead, Julia. *In Colombia, US Companies Get Down to Business* [online], 5 July 2002. <www.americas.org/>.

Palast, Greg. *The Best Democracy Money Can Buy: An Investigative Reporter Exposes the Truth about Globalization, Corporate Cons, and High Finance Fraudsters*. Pluto, 2002.

Parrott, Nicholas, and Terry Marsden. *The Real Green Revolution: Organic and Agroecological Farming in the South*. Greenpeace, 2002.

Partridge, Ernest. *Perilous Optimism* [online]. <http://gadfly.igc.org/papers/cornuc.htm>.

Pelletiere, Stephen C. *Iraq and the International Oil System: Why America Went to War in the Gulf*. Praeger, 2001.

Phillips, William G. "Are We Really Running Out of Oil?" *Popular Science*, May 2000, p.56.

Pilger, John. *Hidden Agendas*. The New Press, 1998.

Pimentel, David et al. "Ecology of Increasing Disease: Population growth and environmental degradation." *Bioscience*, Vol. 48, No. 10, October 1998, pp. 817-27.

Pollan, Michael. "When a Crop Becomes King." *New York Times*, 19 July 2002.

Ponting, Clive. *A Green History of the World: The Environment and the Collapse of Great Civilizations*. Penguin, 1993.

Ponting, Clive. *The 20th Century: A World History*. Henry Holt, 1998.

Rashid, Ahmed. *Taliban: Militant Islam, Oil, and Fundamentalism in Central Asia*. Yale University Press, 2000.

Ray, Paul H., and Sherry Ruth Anderson. *The Cultural Creatives: How 50 Million People Are Changing the World*. Three Rivers Press, 2001.

Rifkin, Jeremy. *Beyond Beef: The Rise and Fall of the Cattle Culture*. Dutton, 1992.

Rifkin, Jeremy, with Ted Howard. *Entropy: A New World View*. Bantam, 1981.

Roseland, Mark. *Eco-City Dimensions: Healthy Communities, Healthy Planet*. New Society, 2002.

Roseland, Mark. *Toward Sustainable Communities: Resources for Citizens and Their Governments*. New Society, 2002.

Rosenberg, Marshall B. *Nonviolent Communication: A Language of Compassion*. PuddleDancer Press, 1999.

Ross, David. *Power from the Waves*. Oxford University Press, 1995.

Salomon, Thierry, and Stephane Bedel. *The Energy Saving House: A Practical Guide to Saving Energy and Saving Money in the Home*. New Society, 2002.

Schoofs, Mark. "AIDS: The Agony of Africa." *The Village Voice* [online], 3-9 November 1999. <www.villagevoice.com/specials/africa/>.

Schweizer, Peter. *Victory: The Reagan Administration's Secret Strategy that Hastened the Collapse of the Soviet Union*. Atlantic Monthly, 1996.

Seo, Danny. *Be the Difference: A Beginner's Guide to Changing the World*. New Society, 2001.

Shuman, Michael H. *Going Local: Creating Self-Reliant Communities in a Global Age*. Routledge, 2000.

Soule, D. D. and Judy D., with Jon K. Piper. *Farming in Nature's Image: An Ecological Approach to Agriculture*. Island Press, 1992.

Smil, Vaclav. *Enriching the Earth: Fritz Haber, Carl Bosch, and the Transformation of World Food Production*. MIT Press, 2001.

Stauber, John C., and Sheldon Rampton. *Toxic Sludge Is Good for You: Lies, Damn Lies, and the Public Relations Industry*. Common Courage, 1995.

Stein, Matthew. *When Technology Fails: A Manual for Self-Reliance and Planetary Survival*. Clear Light, 2000.

Stipp, David. "The Coming Hydrogen Economy," *Fortune* [online], 12 November 2001. <www.business2.com/articles/mag/print/0,1643,34966,00.html>.

Stuart, Paul. *Camp Bondsteel and America's plans to control Caspian oil* [online], 29 April 2002. <www.wsws.org/articles/2002/apr2002/oil-a29.shtml>.

Sussman, Art. *Dr. Art's Guide to Planet Earth*. Chelsea Green, 2000.

Suzuki, David, and Peter Knudtson. *Wisdom of the Elders: Sacred Native Stories of Nature*. Bantam, 1992.

Swindell, Gary S. "Texas Production Data Show Rapid Gas Depletion." *Oil & Gas Journal* [online], 21 June 1999. <http://oil.server4.com/tx-depl.htm>.

Tainter, Joseph. *The Collapse of Complex Societies*. Cambridge University Press, 1988.

Tainter, Joseph. "Complexity, Problem Solving, and Sustainable Societies," in Robert Costanza et al., eds. *Getting Down to Earth: Practical Applications of Ecological Economics* [online]. Island Press, 1996. <http://dieoff.com/page134.htm>.

Tickell, Joshua, and Kaia Roman. *From the Fryer to the Fuel Tank: The Complete Guide to Using Vegetable Oil As an Alternative Fuel*. Greenteach, 2000.

Trainer, F. E. "Can Renewable Energy Sources Sustain Affluent Society?" *Energy Policy*, Vol. 23, No. 12, 1995, pp. 1009-25.

Udall, Randy, and Steve Andrews. *Natural Gas Primer* [online]. <www.oilcrisis.com/gas/primer/>.

Vallette Jim, and Daphne Wysham. *Enron's Pawns: How Public Institutions Bankrolled Enron's Globalization Game* [online]. Institute for Policy Studies Report, March 2002. < www.seen.org/pages/reports.shtml>.

Wachtel, Paul. *Poverty of Affluence: A Psychological Study of the American Way of Life*. Free Press, 1983.

Wackernagel, Mathis, and William Rees. *Our Ecological Footprint: Reducing Human Impact on the Earth*. New Society, 1996.

Wakefield, Julie. "Boys Won't Be Boys." *New Scientist*, 29 June 2002, pp. 42-5.

Werner, David. *Where There Is No Doctor: A Village Health Care Handbook*. Hesperian Foundation, 1992.

White, Leslie A. *The Science of Culture: A Study of Man and Civilization*. Farrar, Straus & Co., 1949.

White, Leslie A.. *The Evolution of Culture*. McGraw-Hill, 1959.

Yergin, Daniel. *The Prize: The Epic Quest for Oil, Money and Power*. Touchstone, 1993.

Youngquist, Walter. *Geodestinies: The Inevitable Control of Earth Resources over Nations and Individuals*. National, 1997.

Youngquist, Walter. "The Post-Petroleum Paradigm — and Population," *Population and Environment: A Journal of Interdisciplinary Studies*, Vol. 20, No. 4, 1999, pp. 297-315.

Zinn, Howard. *A People's History of the United States*. Harper, 2000.

Index

About the Author

Richard Heinberg is a journalist, educator, lecturer, and musician. He has lectured widely, appearing on national radio and television in five countries, and is the author of four previous books, including the original edition of *The Party's Over*, and the follow-up volume *Powerdown: Options and Actions for a Post Carbon Future*, as well as *Cloning the Buddha: The Moral Impact of Biotechnology* and *A New Covenant with Nature: Notes on the End of Civilization and the Renewal of Culture*. The latter was a recipient of the "Books to Live By" award of *Body/Mind/Spirit* magazine.

His monthly *MuseLetter* was nominated in 1994 by *Utne Reader* for an Alternative Press Award and has been included in Utne's annual list of Best Alternative Newsletters.

Heinberg is a member of the Core Faculty of New College of California, where he teaches courses on "Energy and Society" and "Culture, Ecology, and Sustainable Community." He is also an accomplished violinist. He and his wife, Janet Barocco, live in a suburban home they have renovated for energy efficiency, where they grow much of their own food.

Richard's website is: <www.museletter.com>.